THE FASCINATION
OF TIME

Harry Niemann

THE FASCINATION OF TIME

Classic Watches: The Brands, History, and Complications

Schiffer Publishing Ltd

4880 Lower Valley Road • Atglen, PA 19310

Copyright © 2014 by Schiffer Publishing Ltd.

Library of Congress Control Number: 2014951770

All rights reserved. No part of this work may be reproduced or used in any form or by any means—graphic, electronic, or mechanical, including photocopying or information storage and retrieval systems—without written permission from the publisher.

The scanning, uploading, and distribution of this book or any part thereof via the Internet or via any other means without the permission of the publisher is illegal and punishable by law. Please purchase only authorized editions and do not participate in or encourage the electronic piracy of copyrighted materials.
"Schiffer," "Schiffer Publishing, Ltd. & Design," and the "Design of pen and inkwell" are registered trademarks of Schiffer Publishing, Ltd.

Designed by Molly Shields
Type set in ItcSymbolMedFO/Helvetica Neue LT Pro

ISBN: 978-0-7643-4685-9
Printed in China

Published by Schiffer Publishing, Ltd.
4880 Lower Valley Road
Atglen, PA 19310
Phone: (610) 593-1777; Fax: (610) 593-2002
E-mail: Info@schifferbooks.com

For our complete selection of fine books on this and related subjects, please visit our website at www.schifferbooks.com. You may also write for a free catalog.

This book may be purchased from the publisher. Please try your bookstore first.

We are always looking for people to write books on new and related subjects. If you have an idea for a book, please contact us at proposals@schifferbooks.com.

Schiffer Publishing's titles are available at special discounts for bulk purchases for sales promotions or premiums. Special editions, including personalized covers, corporate imprints, and excerpts can be created in large quantities for special needs. For more information, contact the publisher.

Originally published as FASZINATION ZEIT by Delius, Klasing & Co. KG, Bielefeld © Delius, Klasing & Co. KG. Translated from the German by Omicron Language Solutions

Contents

Introduction
What You Should Know about Mechanical Watches 9

History and Museums
The Genius Loci—The Importance of Tradition and the Glashütte Watch Museum . . 15

History of the Watch—The Patek Philippe Museum 23

Auto Museums and Watches—The Montre Classic Collection 31

The IWC—A Company and Company Museum in Schaffhausen 37

From Alpha to Omega—
The Omega Museum. 47

Brands
There's Only One at the Top—Patek Philippe, the Watchmaker 57

Maurice Lacroix—The Way to In-House Watchmaking 63

Every Movement Gets a Crown—Rolex, the World's Largest Luxury Watchmaker. . . . 67

The Rolex Oyster—The Long Road to an Elegant Sports Watch 79

Watches since 1735—Blancpain, a Great Marketing Story 89

From 1775 to Today—Vacheron Constantin, the Oldest Watch Brand in the World. . . . 95

Fascinated by Speed—TAG Heuer, the Racing Driver's Watch 101

Paul Picot—Even a Young Brand Can Make Traditional Watches 105

Now also at the Leading Edge in Marketing Technique—Zenith. 109

Chronoswiss and Rüdiger Lang—
Authentic Feel through the Fascination of
Mechanics 113

The Swatch Group—Combined Luxury
Brands 119

Wearing Time in Fine Style—
A. Lange & Söhne 123

The Fine Craft Workshop in the Müglitztal:
How Are A. Lange & Sohne Watches
Created?. 129

Audemars Piguet—
The Mythos of a Great Brand 137

Jaeger-LeCoultre—Fine Workshop and
Master of Complications. The Movement
Specialist from the Vallée de Joux. 143

Complications and Variations

The Pilot's Watch—
Not Just for Aircraft Pilots 149

The Regulator—
Precision is the Highest Precept 155

The Chronograph—The Sportsman among
Watches 159

The Moon Phase—
The Moon's Disc on Your Wrist 167

The Calendar—
Day and Month on the Dial 173

The Alarm—Even Small Watches Can Ring
Out Loud. 177

The World Time Watch—Time Zones in
GMT And/or UTC. 181

The Tourbillon—Watchmaking's French
Revolution. 185

The Diving Watch—The Only Really Robust
Everyday Watch 193

The Big Date—The Date at Center Stage 199

The Grande Complication—Nothing Less
than What's Technically Doable 203

The Repeater—
Gentle Tones from a Watch. 207

The Perpetual Calendar—A Calendar
There's No Need to Set. 211

The Retrograde Display—Play Instinct with
Surprises 217

Vintage Pieces for Your Wrist

Wearing and Collecting
Historical Watches 223

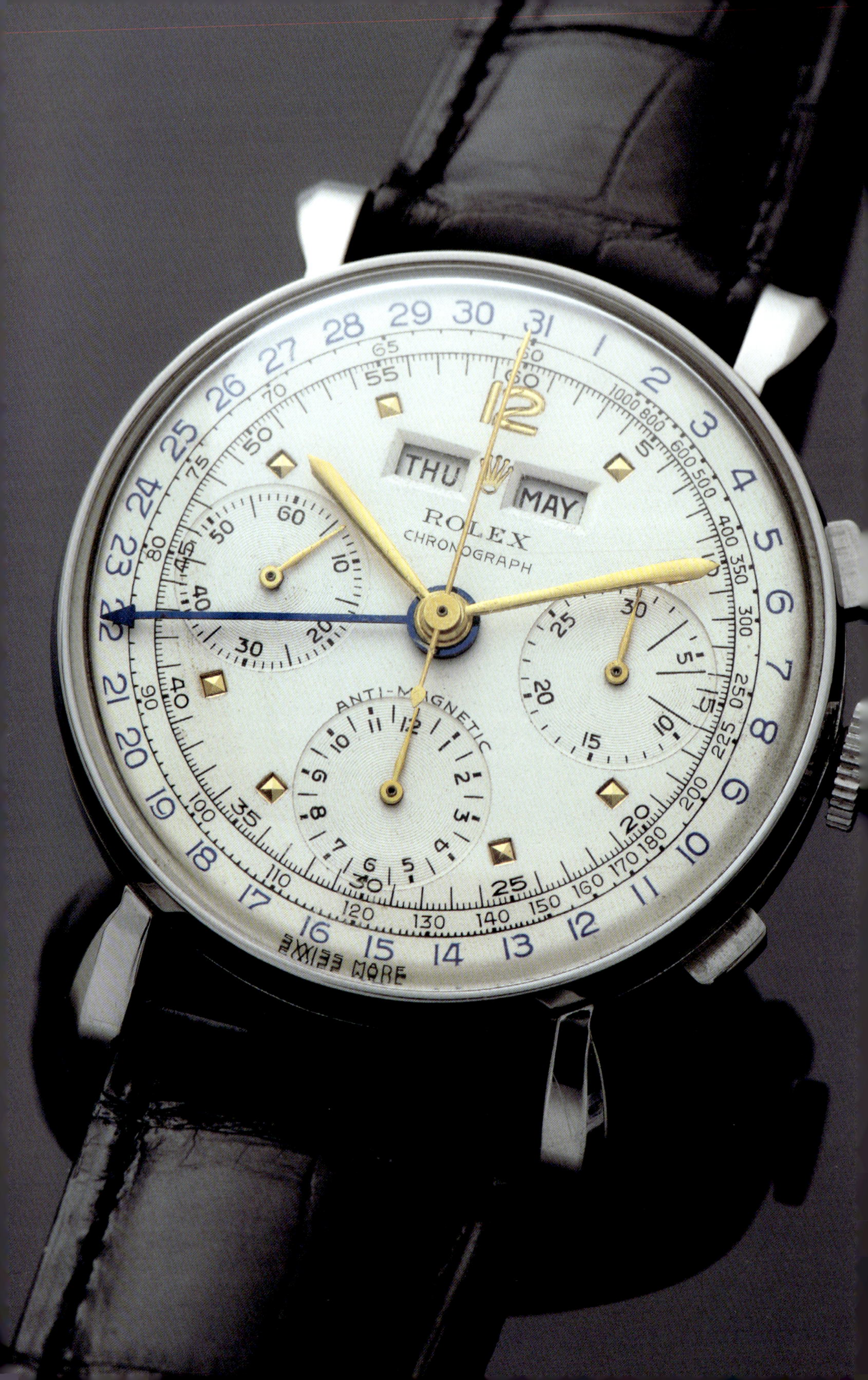
THU
MAY
ROLEX
CHRONOGRAPH
ANTI-MAGNETIC
SWISS MADE

Introduction

What You Should Know about Mechanical Watches

More and more people are succumbing to the fascination exerted by a mechanical watch, although, essentially, it is an anachronism. No other era has showcased the informational value of time so extensively as our own. Telephones, radios, computers, and mobile phones all have a time display. Everywhere, we are surrounded by clocks. Higher-end cars often include clocks as standard features twice over, in both analog and digital display. If you still don't want to give up on wearing a watch on your wrist, the quartz watch offers the greatest accuracy. If that still isn't enough, go for a radio-controlled watch, which will lose only one second in a million years.

Despite all this, or perhaps because of it, the mechanical watch, whether as an automatic or, in its purest form, manual-winding, is experiencing a renaissance never thought possible. In a fast-paced time, when TVs or computers are considered outmoded in just a few years, or often even a few months, we apparently are yearning for things that not only defy the passage of time, but even maintain their initial value. It gives us even greater delight when that value increases.

Anyone who had bought a Rolex Daytona for $800 in 1978, will today get twenty times that contemporary purchase price for it. The same applies to a Patek Philippe Reference 1562. Whoever bought

This calendar chronograph Reference 4768 is one of the most sought-after vintage Rolex models with a Valjoux caliber 72 C; only 220 pieces were made. The date display is by central date hand; there are two windows for day and month.

this manual-winding model with perpetual calendar in 1952, might—provided it is still in good condition—obtain nearly a hundred times the original price today. Such examples give us hope and encourage us to purchase mechanical watches.

These marvels of technology give us much more daily gratification than any stock or bank statement. The universe of the mechanical wristwatch is rich with complications and makers. The great watch brands are a bit of that extra something in life. Even for someone who has collected original watches for years, all featuring the widest range of functions, the time comes when the desire to possess one of the highest quality brands is overpowering. Even the most sober buyer cannot evade the fascination of a continuous history of over 100 years, which gives the product a patina that raises these watches to the technical nobility.

Patek Philippe, Audemars Piguet, Blancpain, Breguet, Rolex, IWC, Omega, Zenith, Vacheron Constantin, and A. Lange & Söhne are some of these brands. In every decade, they were members of the *Haute Horlogerie* and true masters of complications. The tourbillon with chronograph, the minute repeater, as well as such exotic complications as a Grande Sonnerie Carillon, or even a perpetual calendar with world time display and a perpetual calendar with sunrise and sunset and equation of time—all watches, some costing well over $65,000 and thus for most of us something we could only ever dream of owning, that formed the *oeuvre* of these manufacturers. It is legitimate for such companies to invoke the magic of tradition and, mindful of these magnificent designs of the past, to create these anew, and in even more sophisticated forms using modern production technology and design computers. What would watchmakers of past generations have said, if you had shown them a waterproof minute repeater, like that offered by Blancpain?

Anyone who has once succumbed to the fascination of the mechanical watch, will not usually let things go with the purchase of just one such watch. Perhaps they will collect according to brand or complications, and/or watches with a certified chronometer. In doing so, a chronometer is often confused with a chronograph, or, analogous to the word meaning, is understood to be a simple watch. In addition, not every certified chronometer is a watch with a mechanical movement. What counts for the lovers of mechanisms, is that a mechanical watch

A watchmaker can only use the term Manufacturer—in-house workshop-made—in the real sense, if it also manufactures its own movements. Here is a convincing example of proprietary or in-house watch movement making by Jaeger-LeCoultre.

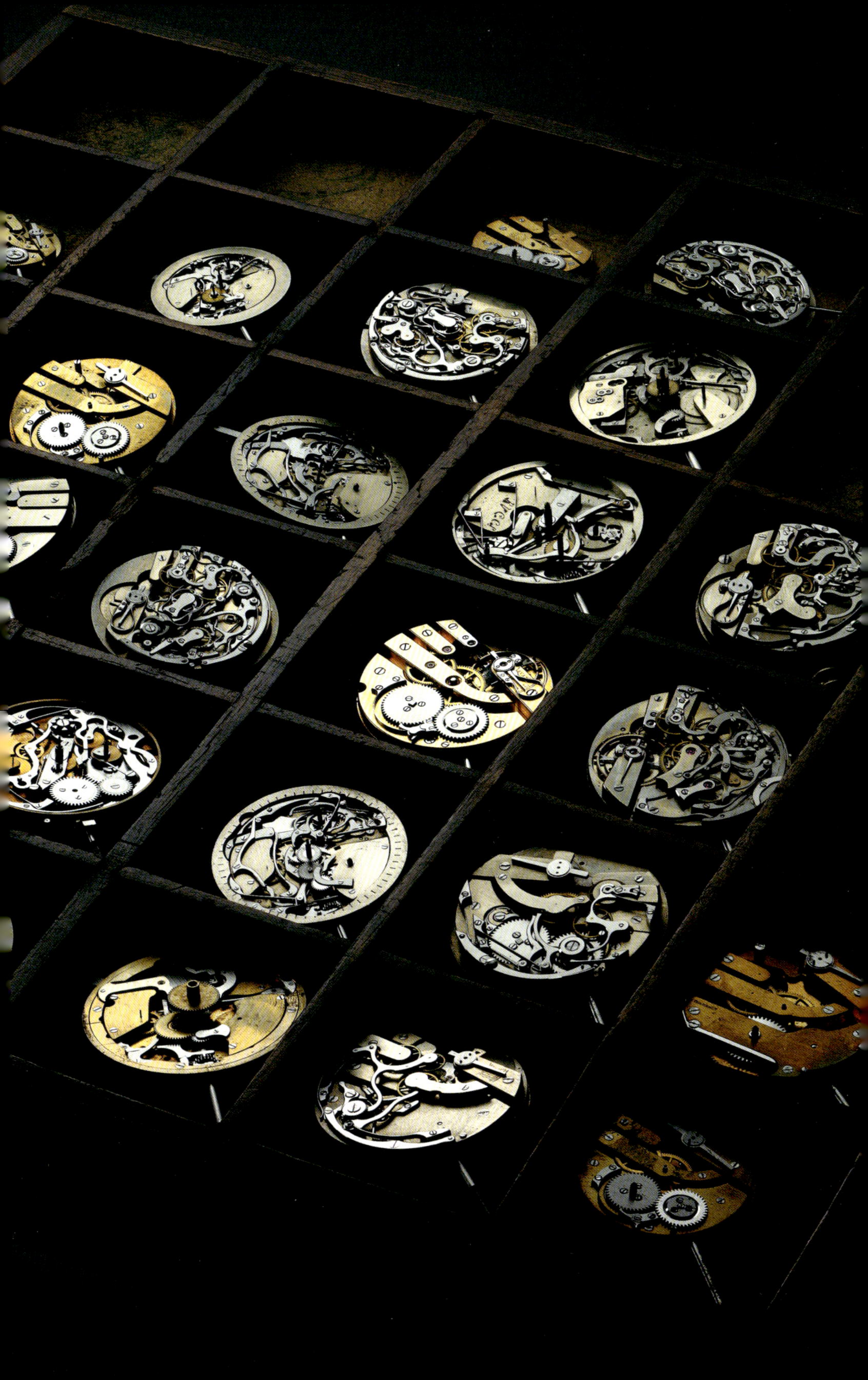

The IWC Da Vinci, now in tonneau shape, with perpetual calendar and chronograph. It displays up to the eighth of a second, as well as the sequence of the months and leap years.

You will find the most important complications in this book. Beyond that, the book discusses brands, museums, and complications, although with no claim to completeness. For those interested in watches, who are toying with the idea of purchasing a high-quality mechanical watch, the basic knowledge of what you should know when buying a mechanical watch is offered. Ultimately, however, a mechanical watch is more a fashion accessory than a reliable timepiece. This task is fulfilled much better by a radio-controlled watch that gets its time signal accurately from an atomic clock, and even by a high quality quartz watch which loses just one second a month—a precision that a mechanical watch attains just once in 24 hours.

Anyone who collects watches, can indulge their fashion play instinct. Combine watches of yellow gold, white gold, rose gold, or platinum and steel with the matching cufflinks and a signet ring of the same metal; coordinate the leather strap with belt and shoes and wear your sports watch where it belongs. No diver, not even a diver for treasure, wears a gold watch under water, even if it allows a depth of 2,000 meters. A complication is better suited to an elegant case than as part of a sports watch. If you still don't want to do without precious metals for sport and leisure, there is the white gold watch, with white gold or rubber band, as a stylish alternative. Of course, white metals such as white gold, steel, and platinum can also combine well with each other. In addition to the current models, the now-over-100-year

which is certified as a chronometer, is one which has been tested for precision by the COSC (*Contrôle Officiel Suisse des Chronomètres*). Even if it is merely a snapshot, watch movements with this test certificate are very finely adjusted. Omega was one of the biggest chronometer manufacturers in the 1950s and 1960s. Today, it is the Rolex company which produces the largest share of mechanical wristwatch chronometers.

history of the wristwatch invites you to wear a genuine vintage model on your arm. Indulge yourself at the many watch auctions or take a glance through their catalogs, and discover your favorite bygone eras. Often, equipped with a little expertise, you can obtain some interesting bargains. A recent example is the Ebel E-Type, which was only made for a relatively short time starting in 2001. Today, this watch, a chronometer-chronograph with the finely finished Caliber 137, is often available as a never-worn stock item for a third of the original price. It is the more surprising that Ebel discontinued this model line, because this type launched the trend for oversized watches.

Often, an older watch has a special charm, especially when it, at first inspection, exactly matches the current fashion trend. You get a real surprise when you learn by asking, that it already existed 35 years ago. Not least, this also explains the exorbitant and objectively unjustifiable prices which vintage Panerai models attain at auctions, since in this case the historical model was used to design the new line of watches. In contrast to the past, no one needs a clamp over the crown to seal it watertight today, and yet such anachronistic details are savored by customers. Vintage Rolex models are always eye-catchers; their patina alone conveys a different aura than any new watch. This is how you can not only get to know and master the history of watches, but also wear it.

The Tourbograph is one of the most complicated watches made by A. Lange & Söhn; it has a fusée-and-chai transmission. The complications offered are a tourbillon and a rattrapante or double chronograph.

Deutsches
Uhrenmuseum
Glashütte
Glashütte
ORIGINAL
UHRMACHERSCHULE
Alfred Helwig

History and Museums

The Genius Loci—The Importance of Tradition and the Glashütte Watch Museum

We live in a time when product and company traditions mean more than just simply preserving the past. Traditions and caring for those traditions contribute decisively to shaping a product's image.

The product's historical continuity itself is a hallmark of excellence, and even more so if it had a real influence on a particular epoch. Tradition thus becomes a selling point. Of course, there is more to history: as the philosopher Hermann Luebbe once put it, it has a lot to do with our identity. In a largely uniform and yet also anachronistic period of watchmaking, a product's history gives it its necessary identity. It is not for nothing that some manufacturers try to provide evidence of historical continuity, and then utilize this in an elaborate image campaign. The vintage watch, and this applies to both pocket and wristwatches, undergoes a broad metamorphosis, from being an article for daily use to become a cult object; it thus becomes a new character, outside the sphere of something useful: a new aesthetic dimension, brought to life by craftsmanship and its history.

The fact that companies have been making quality products for more than 100 years, is based not only on professional competence, but also on its spirit. Anyone who can fly to the moon, whether from the East or West, in principle has the ability to manufacture high-technology products. This alone is not enough, however. It is the spirit and mind that determine the nature of a company and its products. And nothing is more important than the care and preservation of this spirit. A company's image is shaped by its people. And where there is a true spirit, the product is also true.

It is therefore no wonder that that the top-level companies in the watch industry invest heavily in the area of preserving their traditions. Visible indications of this are the many museums and museum-style exhibits

The former School of Watchmaking in Glashütte has been turned into an impressive museum; it ingeniously links the local history with watchmaking.

which have sprung up in the recent past. Omega, Longines, Audemars Piguet, IWC, TAG Heuer, Jaeger-LeCoultre, and Patek Philippe have all opened up trademark or factory museums, which include feature activities that go far beyond the mere presentation of vintage watches. A first and important step in this process has been setting up the archives that make it possible to furnish information about the pieces presented and objects exhibited, as well as the quotations cited from the original documents.

We can identify two strategies for these museums: one is the pure company museum, that only features products and brands from its own history; and the other is a museum presentation that documents the history of the mechanical watch, above and beyond the make and company, and portrays each major step in development—regardless of which company ultimately brought this development to the market.

The Deutsche Uhrenmuseum Glashütte—the Glashütte GermanWatch

Silver mines and watchmakers shaped the economic picture in Glashütte, documented here by a Saxon Silbertaler coin and a pocket watch.

Museum—occupies another special position in the diverse museum landscape. Housed in the historical building of the former German School of Watchmaking, the Stiftung Deutsches Uhrenmuseum Glashütte Foundation has created a place that preserves the rich cultural heritage of the city and at the same time promotes watchmaking craftsmanship, training, and science. It is a look at the watchmaking history of a city and region that has significantly influenced the development of the mechanical watch.

The city of Glashütte, with its 500-year history, can look back on over 160 years of turbulent, continuous, and fascinating history of the watch industry. During this time, many

At the end of the tour, the visitor is shown the Glashütte of today in the form of the products of local enterprises: A. Lange & Söhne, Glashütte Original, Mühle, Nomos, and Wempe represent the new watch culture.

important buildings were built for the watch industry. Today, the majority of these buildings either still, or once more, serve for the production of traditional Glashütte timepieces. The major exception remains the building of the former Deutschen Uhrmacherschule – German Watchmaking School in the heart of the city. To give this historically important place a new purpose, in autumn 2005, the city of

One of the special edition Glashütte Original, limited to 130 pieces, made in honor of Karl Moritz Grossmann, who started his watchmaking operation in Glashütte in 1854.

Glashütte announced a contest for proposing innovative concepts for using the building.

As a result of this announcement, the city of Glashütte established a joint foundation with the manufacturer Glashütte Original, which served to establish and operate a watch museum, an archive, a library, as well as demonstration workshops, for the benefit of the public. The city of Glashütte also contributed the watch exhibits in its possession to the foundation.

Glashütte Original acquired—with generous support from The Swatch Group AG—the building of the former Watchmaking School and renovated it completely. In addition, Glashütte Original's Alfred Helwig School of Watchmaking has recently moved into this same building, to carry forward its historical purpose.

Based on the exhibition's motto of "The Fascination of Time—Bringing Time Alive," the innovative Watch Museum not only illuminates the art of high-quality watchmaking itself, but also creates an emotional and philosophical approach to the phenomenon of time. In harmony with the objectives of the German Watchmaking Musem Glashütte—Nicolas G. Hayek Foundation, Glashütte Original has developed a tour through the exhibition which conveys both the rich history of

Why only allow a view of the artistry of the Caliber 66's hand-engraved butterfly bridge through a transparent display back? The PanoInverse XL displays the art of Glashütte watchmaking on the dial.

Glashütte as well as aspects of the sense of time and time measurement.

On two floors and in 1,000 square meters of exhibition space, more than 400 unique exhibits are presented and brought to life using multimedia. Glashütte pocket watches, wristwatches, and clocks of different periods, marine chronometers and escapement models, historical documents and patents, tools and workbenches, as well as astronomical models and metronomes are artfully staged and displayed, Thematically, the exhibition is composed of a series of "history rooms" and "time rooms," framed by a prologue and an epilogue.

The history rooms create the historical context of this watch city; at the entrance are presented its famous people and founding fathers, such as Ferdinand Adolph Lange, Julius Assmann, Adolf Schneider, and Karl Moritz Grossmann, who made Glashütte the traditional stronghold of fine German watchmaking and the training of watchmakers. Along the course of the exhibition, the further epochs that have shaped Glashütte are presented, such as the period of its founding, the First and Second World Wars, the period of dismantling and expropriation, as well as German national reunification and its new foundation. These include displays of the watches produced by the VEB [*Volkseigener Betrieb*, nationally owned Enterprise] Glashütter Uhrenbetriebe (GUB) during the era of the German Democratic Republic.

The time rooms interrupt the chronological sequence of Glashütte's watchmaking history and take visitors into such exhibits as a microcosm of a mechanical watch, to allow them to experience themselves the precision and interplay of its hundreds of individual parts. Another multimedia time room, which includes an interactive glossary of timekeeping, invites you to independent discovery. Glashütte Original refreshingly keeps itself in the background in all the exhibits. In the last room of the museum tour, the visitor can admire a broad range of this and other local brands, such as Lange & Söhne, Nomos, and Mühle. For those who want to delve deeper into the fascinating range of watches shown there and the history of Glashütte Original, there is no getting around a visit to the factory and a tour of the workplace.

The Treasure Chamber, where the three jewels of the art of Glashütte watchmaking can be admired..

"La Grandiose" | Universaluhr
Uhrenfabrik Union | 1899

Universal Watch
by Union | 1899

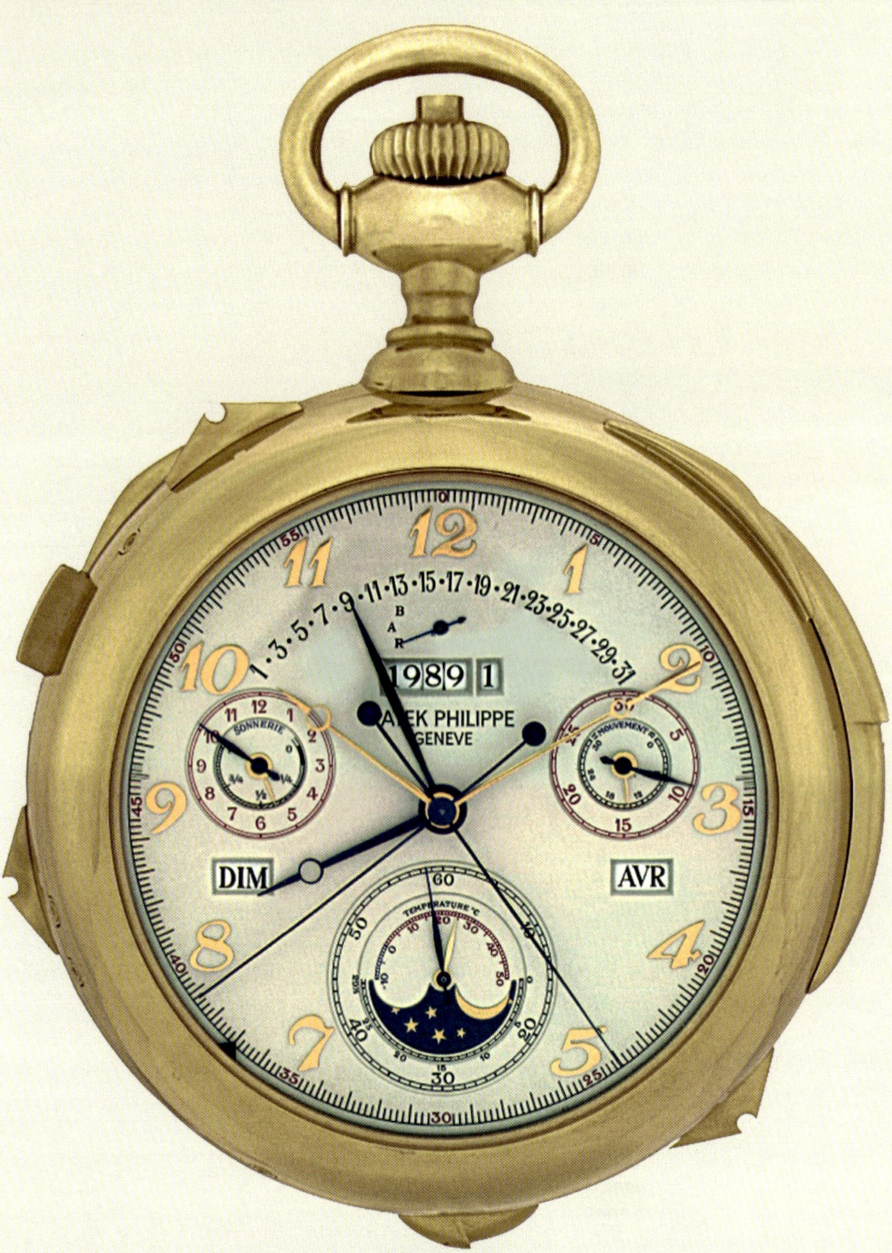
1989
GENEVE
SONNERIE
MOUVEMENT
DIM
AVR
TEMPERATURE °C

History of the Watch—The Patek Philippe Museum

When did the history of the mechanical watch begin? With Peter Henlein of Nuremberg in the 16th century, or was it Jacques Blancpain who started the first watchmaking workshop in the Vallée de Joux in 1735? The oldest watch manufacturer, consistently offering its products on the market since 1775, is the Geneva-based Vacheron Constantin. Technology enthusiasts are generally also interested in the history of technology, in the origin and coming into being of major brands and their inventions. This applies to watches as well as to locomotives, airplanes, or automobiles. Interested observers would like to see the milestones that have given rise to the mythos about product and brand.

Presentation of the artifacts allows the connoisseur to make an enjoyable tour through time. But a museum is, of course, more than just some building, filled with old things. The core is the collection, the basis from which the exhibition is staged. Of course, here as elsewhere, the staging is important. This combination creates the fascination of museums. If, then, the history of your own brand is very closely interwoven with the history of the artifacts themselves, this creates the conditions for a compelling presentation.

Everything stated above applies to the Patek Philippe Museum in Geneva. In the Geneva Plainpalais district, at No. 7, Rue des Vieux-Grenadiers, the Geneva manufacturer houses its treasures and presents them on four floors, each of 700 square meters. American granite for the floors, tiles from Provence for the walls, Alpen serpentine stone for the friezes, Spanish marble for the entrance and stairs—all these materials were carefully selected in regard to their function and colors. Together with the textiles and lighting, they create a calm, relaxed, and inspiring atmosphere for presenting the watches.

The woods used were selected based on special criteria and are as rare as they are fine. Each floor required 1,500 square meters of wood veneer, with uniform color and grain. Four types of wood were used: solid oak for the furniture designs, eucalyptus for the cases for the collection's antique pieces, rare sycamore wood for the collection

The Caliber 89 was made for the 150th anniversary celebrations. It is one of the most complicated mechanical watches, with 33 complications and comprising 1,728 parts. It required five years development time and four years to manufacture it.

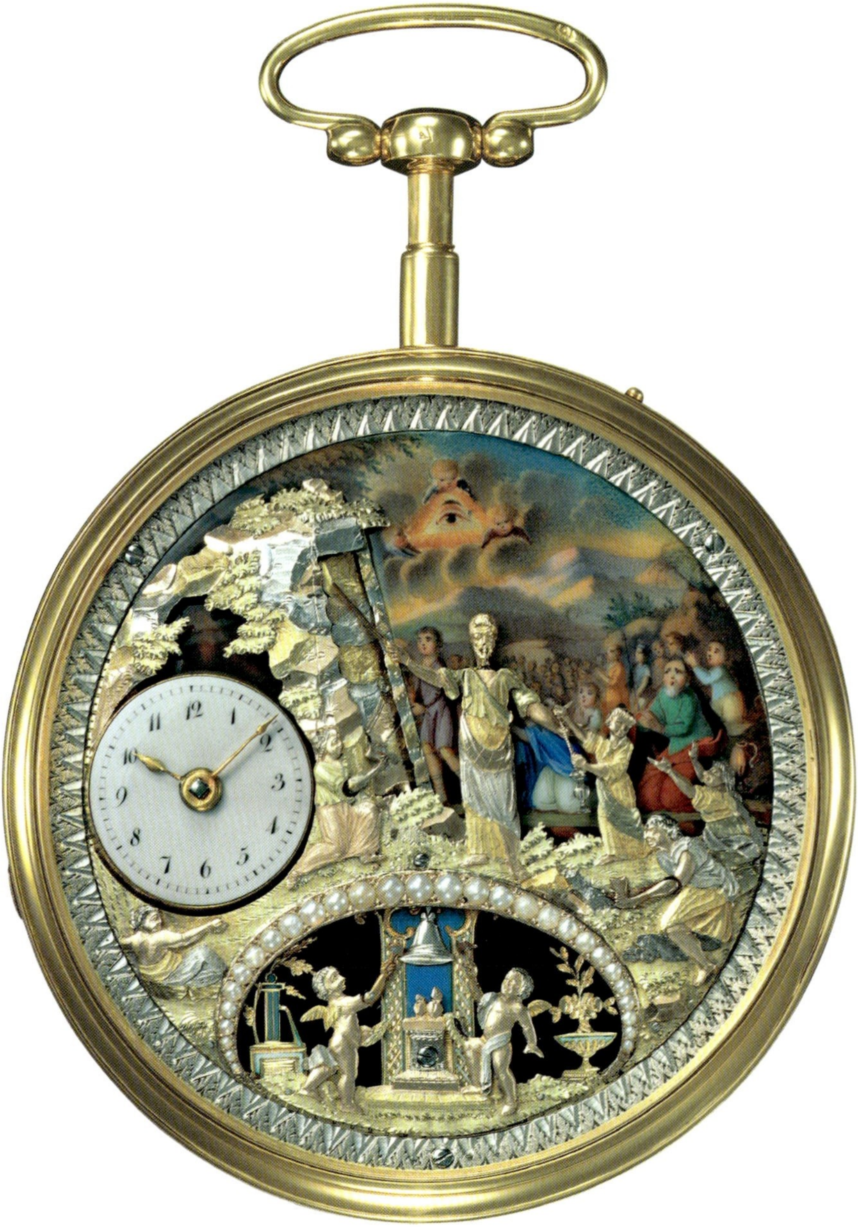

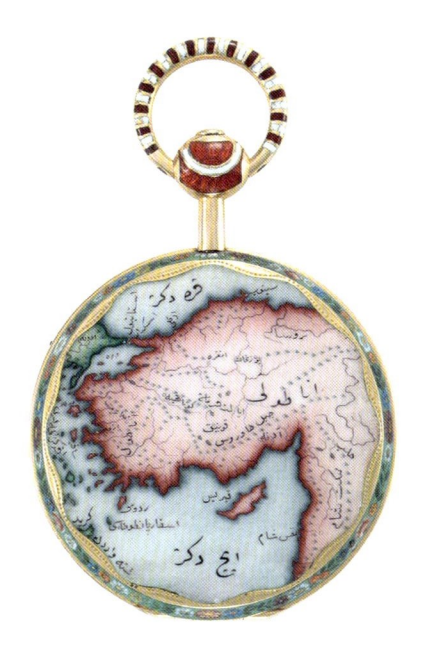

For centuries, pocket watches dominated display of the time for individuals. A watch was, however, also a piece of jewelry and an object of prestige. The Patek Philippe Museum displays such masterpieces as this repeater watch with a Biblical Moses motif, or an enamel watch showing a map of the Ottoman Empire. The Patek Philippe Gondolo with crown winding mechanism has a double 24-hour display, rather unusual for pocket watches.

of Patek Philippe watches, and white oak for the library. The elaborate wood furnishings alone, made from thousands of individual parts, kept some forty carpenters busy for a whole year. Since the House itself is an historically significant building, only conservative measures were undertaken in the renovation, which began in 1999.

The Patek Philippe Museum invites visitors on a tour through four floors; it starts on the ground floor, then moves up to the fourth, down to third floor, and finally ends on the second floor. On the ground floor, next to the reception, there is also a collection of old tools, a watchmaker's workshop, and the auditorium which take visitors back into the early years of watchmaking. An elevator takes you to the fourth floor, which houses the library and the Patek Philippe Archive, the centerpiece of the collection and Museum. The stairs take you down to the third floor, where the collection of antique watches from the 16th to 19th centuries is housed. On the second floor is the area that makes the hearts of those who love this watch maker beat

Médaillon savonnette à tact is the name of this Breguet watch, which allowed the wearer to feel the hours.

fastest: the Patek Philippe collection from 1839 to the present.

The pocket watch has existed for almost 350 years, since the time when Polish Count Antoine Norbert de Patek and his compatriot François Czapek decided, in 1839, to establish a watchmaking firm. After a surprisingly short time, and in large part due to the talented French watchmaker Adrien Philippe joining the partnership in 1845, the Geneva enterprise rapidly gained an excellent reputation, which has kept growing continuously over the 175-year-plus period the firm has been manufacturing watches. The two founders and name-givers of the House of Patek Philippe had vowed to each other that they would make only the best and most beautiful watches in the world. A proud resolution, which then manifested itself in one of the first keyless watches manufactured.

Now the Patek Philippe Museum, with over 1,000 watches, tells the story of one

The Cross of the Order of the Holy Spirit is one of the watches from the early 17th century (left).
The view across Lake Geneva to Mont Blanc was a popular motif for artists in enamel. Here, a pocket watch intended for the Chinese market (right).

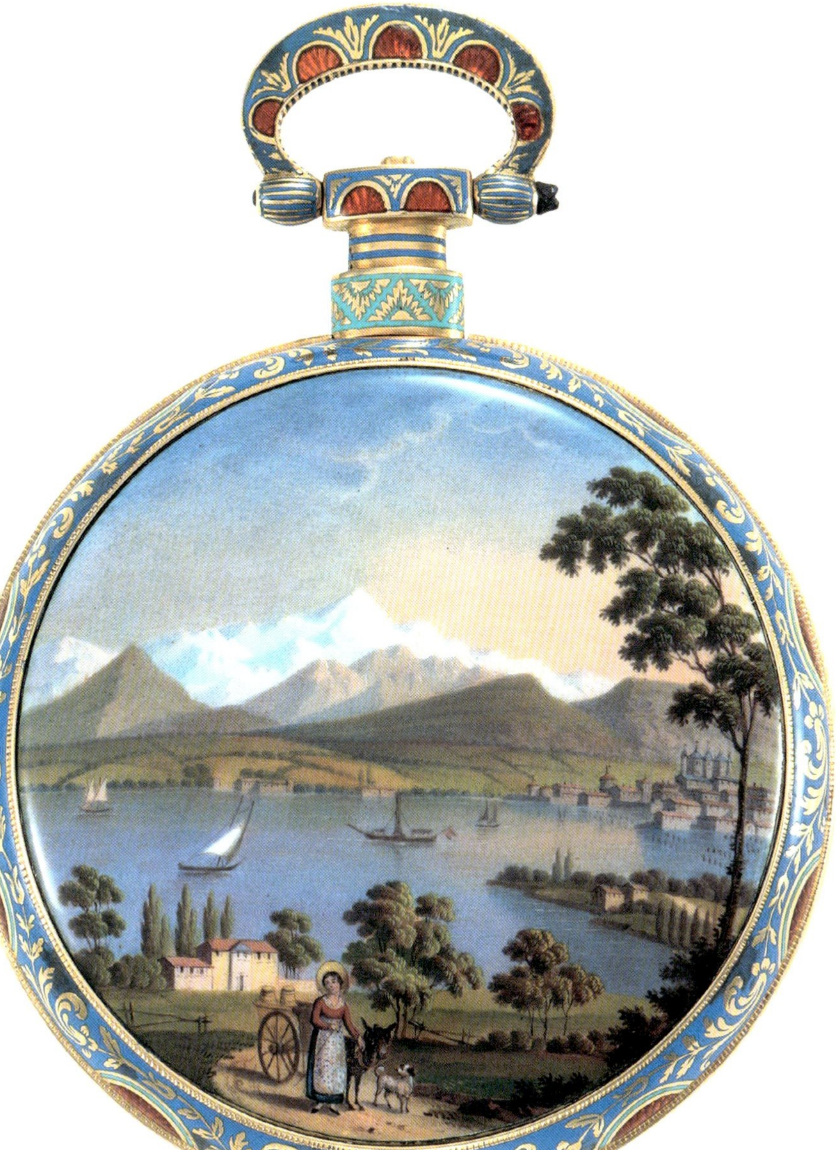

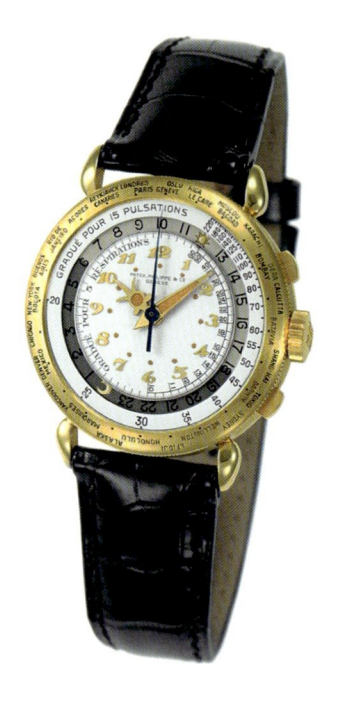

Calatrava men's watch Reference 570, with calendar and moon phase display. Patek Philippe world time watch (right).

of the most creative manufacturers of our time, whose innovative energy, both technical and in design, is as strong as ever. Everything in watch design, from Art Nouveau to Art Deco, is on display, as well as the such special technical features as world time watches or special model series, such as Calatrava wristwatches that, first launched in 1932, very quickly set the standard for watchmaking elegance. The approximately 500 different wristwatches alone delight aficionados of this brand, yet the watches with complications can top even that experience. Both the myth-shrouded New York banker Henry Graves Jr. and legendary car designer James Ward Packard had placed an order with Patek Philippe to make a pocket watch with the most state-of-the-art complications. Thus, over a few years' interval, the Geneva watchmaker created the two most complicated wearable watches in the world. The Graves piece was finished in

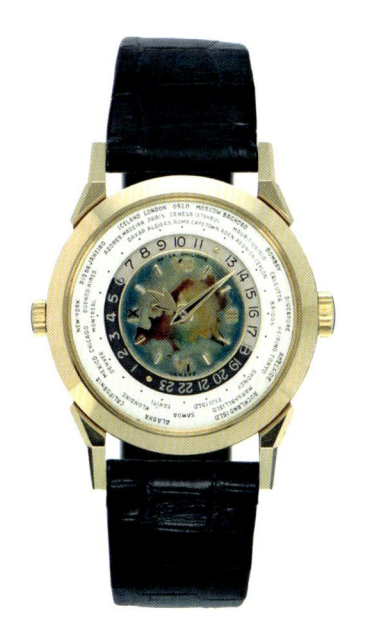

Rare split-seconds chronograph by Patek Philippe with small seconds, 30-minute totalizer at the three and a calendar with moon phase display; next to it, a world time watch with enameled dial.

1933, after six years of work, and, with its 24 complications, surpassed even the 1916 Packard watch, which already had united 16 complications in a single watch movement.

The collection, which makes such an exhibition possible, was compiled by Alan Banbery, a personal adviser to Philippe Stern.

The Patek Philippe Museum is open to individual visitors and group tours from Tuesday to Friday, from 2 p.m. to 5 p.m., and Saturday from 10 a.m.10 to 5 p.m.. It invites you to take a fantastic journey through time, starting with a *Dosenuhr,* or drum-shaped watch, (ca. 1530) and then leads you to the impressive automaton Moses (ca. 1820), to the most complicated pocket watch in the world (1989). The only disadvantage: there are too few watches from Patek Philippe's current model range in the museum. The likely reason is that many clients worldwide are desperately waiting for their watches, and therefore they can in no way show them in the museum.

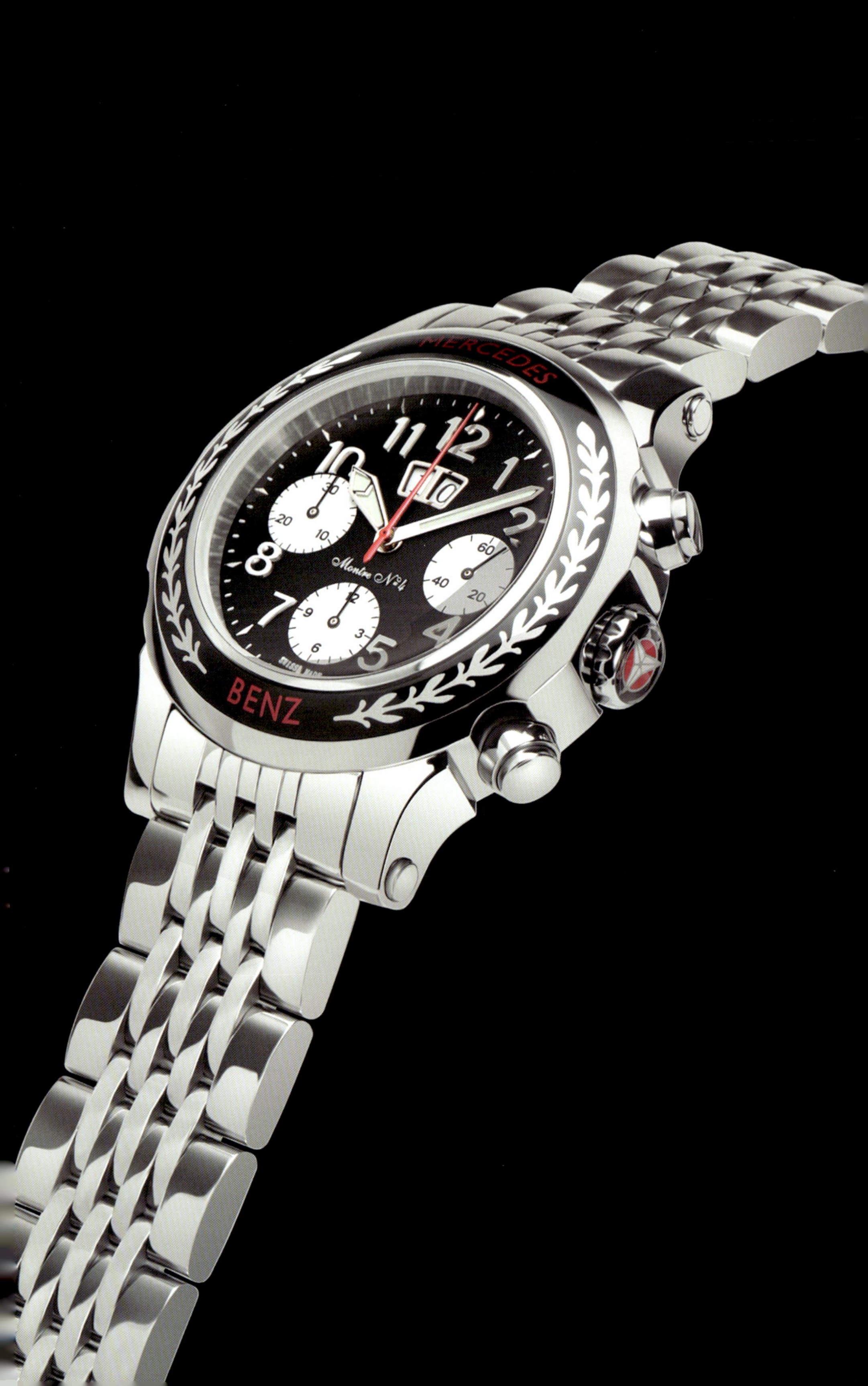
MERCEDES
BENZ
Montre N°4

Auto Museums and Watches—
The Montre Classic Collection

Watches and museums are anchors. They help us handle the speed in the world. At Mercedes-Benz, the Museum itself is a tradition. The first small exhibition of historical vehicles was in 1923, and, right on time for the 50th anniversary of the automobile—which Karl Benz had invented in Mannheim in 1886—the first real museum was opened, in 1936. However, things also took place in Stuttgart. In 1883, Gottlieb Daimler and Wilhelm Maybach, by inventing the first high-speed gasoline engine, created the foundation for the level of mobility we have today. Three years later, at the same time as Benz, these two brilliant inventors also taught the automobile how to drive. And in 1904, not far from the city where these historical milestones are, the Daimler Motor Company factory was founded in Stuttgart-Untertürkheim. The current, new since May 2006, Mercedes-Benz Museum is therefore located directly next to the cradle of the automobile—could any site be more fitting than this?

And what could be more fitting for the opening of such a museum than a watch, specially designed for the occasion?

The Montre Classic Chronograph No. 4 is a sequel to the extremely successful watch

On the Montre No. 4, with big date and chronograph, the watch bracelet can be changed without any tools, because of the quick-release fasteners on the lugs.

Both futuristic and imposing: the Mercedes-Benz Museum. Here you can experience automotive history, using the example of one make of car.

1945
1960

Form und Design
Form and design

The Mythos Room 4, where the fascinating super sports cars of the 1950s and 1960s, such as the 300 SL gull-wing door or the 300 SL Roadster, are shown.

The Montre No. 4 crown picks up on a red version of the Mercedes star, as used by the Daimler Motor Company.

creations by the House of Daimler, and, like its predecessors, was developed according to very specific criteria and specifications. Sportsmanship and tradition, always combined with the latest technology, have always been the defining qualities of the Montre Classic Collection for the company with the star.

Now, we already have a fourth model of the highly successful series of watches in the Mercedes-Benz Classic range. All these watches are designed in the form of either a radiator or radiator cap. The bezel features the bay leaf, with the inscription "Mercedes Benz." Numbers 1 and 2 were chronographs, first equipped with a Valjoux 7750, the second time with a Lemania 1882.

The new automatic chronograph is equipped with a refined version of the exclusive Caliber Dubois Dépraz 4500. Finely-crafted decoration—so-called Geneva perlage or circular graining—distinguishes this version along with oscillation and escapement mechanisms of the finest quality from the basic movement.

The basic caliber for this modular movement is the ETA 2892-2, with 28,800 vph and 42-hour power reserve when fully wound. The big date is set below the twelve. The totalizer for the 30-minute display is located at the nine; the 12-hour display at the six. The small seconds is at the three. The hour markers are appliqué and so

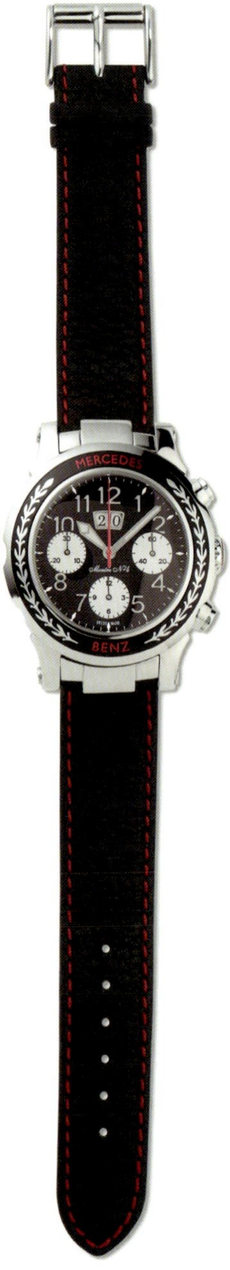
MERCEDES
BENZ

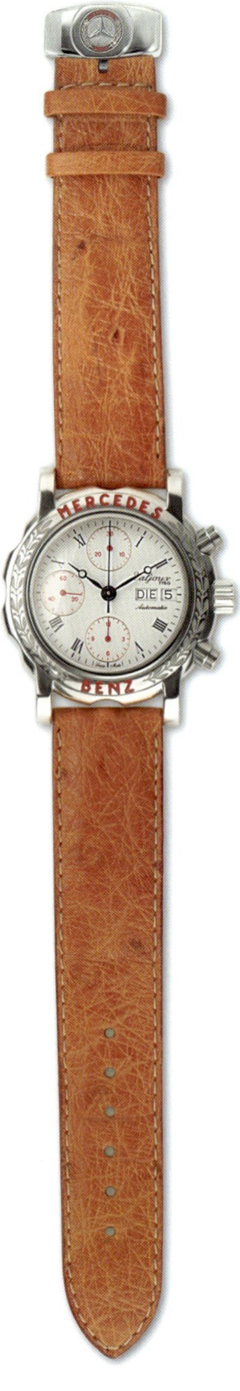
MERCEDES
BENZ

executed that, together with the luminous lance-shaped hands, they ensure optimal readability. The movement has a stop seconds mechanism, so that the time can be set precisely,

The discs for the big date are on one level, but separated by a divider which has the task of covering the gap between the discs.

The date switch-over gains about a quarter of an hour at midnight. Of course, there is a date quick-set mechanism, which makes it possible to set the watch quickly after it has not been worn for some time. A sapphire crystal, in the shape of the ground plan of the new Mercedes-Benz Museum, is set in the screw-down case back, providing a view of the rotor and movement, which is decorated with an appealing Geneva perlage.

The watch has an anti-reflective sapphire crystal. The crown is another eye-catcher, decorated with a classic Mercedes logo in red. The chronograph push-pieces are round. On this watch, you can change the bracelet without using any tools. Two push-pieces on each case lug unlock the bracelet, letting you quickly change from a leather strap to a steel bracelet. The watch comes with both. The Mercedes logo is etched on the metal bracelet buckle. The leather strap buckle is modeled on an old car hood latch. The watch comes in two versions, as the Set Classic and Set Sport, with cream colored dial, black totalizers, date display, and a black dial with white indicators. The bezel on the black version is also black, with red lettering. On the light-colored version, inscription and laurel wreath logo are engraved. The black calf leather straps for both are stitched in either white or red, matching the bezel. The watch comes in an elegant black piano lacquer box.

Only 333 of these watches have been made. Using the 30-minute and 12-hour counters, the Montre Classic Chronograph No. 4 clocks time intervals precisely and also displays the date via two adjacent discs.

However, the ultimate highlight of this watch is the mechanism for changing the bracelet. Within seconds and without any tools, you can fit either a leather strap or a steel bracelet.

Montre No. 4 and No. 1. Chronographs as ambassadors of the Mercedes-Benz Museum. No. 1 has a Valjoux 7750; the No. 4 has a Dubois Dépraz 4500 modular Caliber.

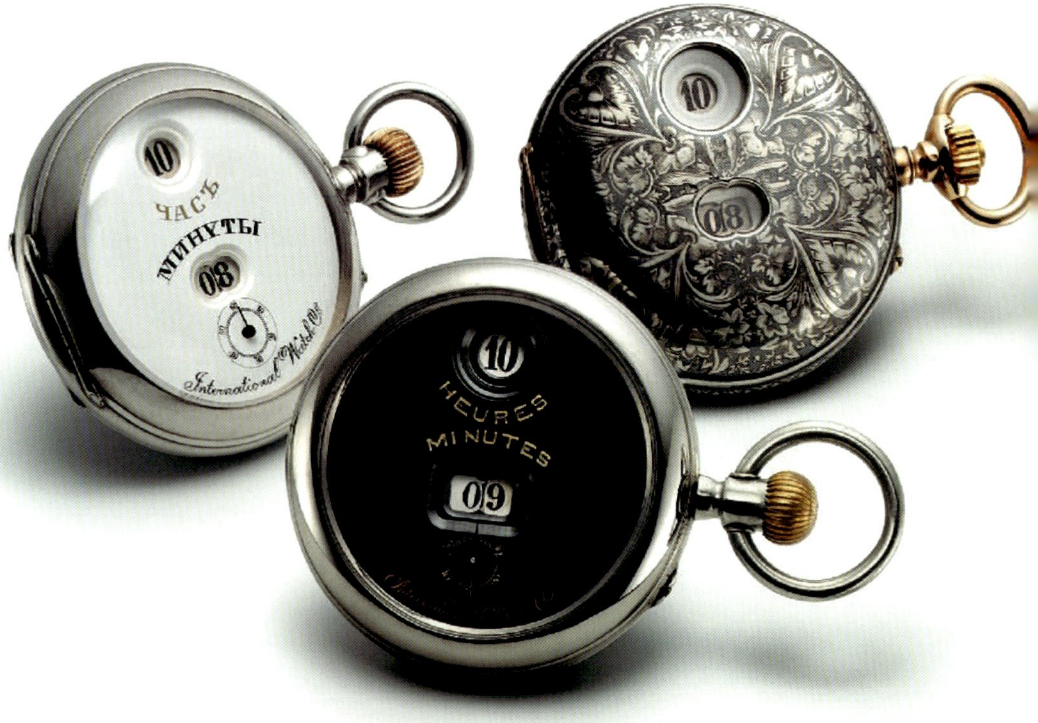
ЧАСЪ
МИНУТЫ
International Watch Co
HEURES
MINUTES

IWC—A Company and Company Museum in Schaffhausen

When the American watchmaker and engineer Florentine Ariosto Jones founded the first and only watch factory in the north-eastern Switzerland in 1868, he did so to supply the expanding American market with high-quality pocket watches. Manufacturing conditions were favorable in Switzerland, so he had the watch movements made there, which were then put into cases in America. This is why early pieces have the inscription "International Watch Co., New York" on the movement bridge.

The new company expanded in Schaffhausen. In 1874-75, a new factory was built in the former orchard of the **Kloster Allerheiligen** (Abbey of All Saints), which remains the IWC's center today. Watch manufacturing began there in the spring of 1875. That year, the IWC had 196 employees. The watchmaker produced pocket watches using the latest machinery. However, the high investment costs for the

The IWC Pallweber, a pure mechanical pocket watch with digital hour and minute displays. The beginnings: the IWC factory building in 1870 (below)

The famous Jones Caliber in an American-made case. The design represents a Mississippi steamboat.

building and machinery led to disputes between the IWC shareholders and F.A. Jones. When production started in the new headquarters, Jones left Switzerland and therefore never witnessed how IWC developed into a world famous watchmaker.

The Schaffhausen Handelsbank provided the necessary cash injection. The company was sold in 1880 to the Schaffhausen machinery manufacturer John Rauschenbach-Vogel, who passed it on to his son. His two daughters both married men who, each in his own way, made a special contribution in their field: the one, C.G. Jung, as psychologist and psychiatrist, the other, Ernst Jacob Homberger, as director and co-owner of IWC. Jung sold his shares to his brother-in-law in 1929, who now became sole owner and left the company to his son Hans Ernst Homberger in 1955.

IWC ended its existence as a family enterprise only when it was sold in 1978 to the instrument manufacturer VDO.

The company had, like many in the industry, pinned its hopes on the quartz watch. IWC was involved in the development of the Beta 21 quartz movement, which runs at a frequency of 8192 hertz. IWC mounted this revolutionary technology in a watch bearing the name of the great genius Da Vinci, since precision was considered the highest precept of watchmaking. But soon the once-great-hope of quartz technology became a nightmare for IWC, like others. It is thanks to former CEO and later board chairman Günther Blümlein, that the company returned to the traditional values of watchmaking. IWC benefited from the fact that, since 1968, watchmakers have been continuously trained to make mechanical movements. IWC was able to position itself in the 1980s with both designer and diver's watches designed by F.A. Porsche, as well as with mechanical masterpieces that were offered at relatively inexpensive prices. One example is the Da Vinci chronograph, introduced in 1985, with a perpetual calendar that is mechanically programmed for 500 years. With this came the correction function using only the crown and a four-digit year

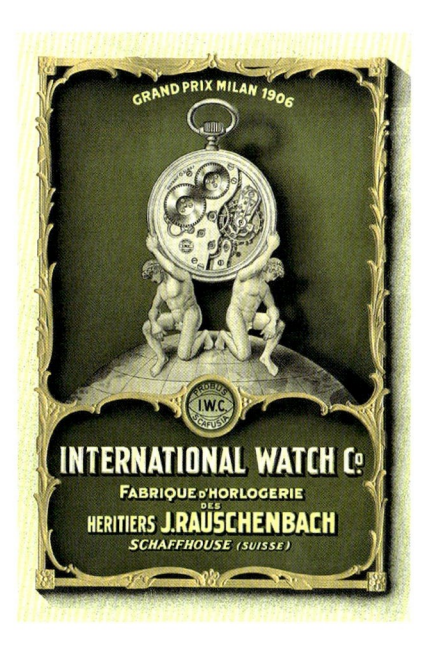

Prizes were used extensively in advertising. Here, a poster from 1906, when the company belonged to the Swiss family Rauschenbach.

display. A year later, this revolutionary watch was offered in a case of zirconium oxide, an absolutely scratch-resistant ceramic. The Novecento is another fascinating model, a rectangular automatic watch with perpetual calendar, in which all functions can be adjusted by the crown and therefore works without any correction push-pieces. This watch is also waterproof, something which until then was very unusual for rectangular watches. They are available in platinum and yellow gold.

But all this was still insufficient for watchmaker Blümlein. In 1990, arrived the Grande Complication for wrist wear. This automatic watch features, besides the perpetual calendar with four-digit year display, a chronograph and moon phase display and minute repeater. IWC required a full seven years development time for this watch.

For the firm's 125th anniversary, a very special watch was designed, with the martial title *Il Destriero Scafusia*. In addition to an extraordinarily beautiful finish on the movement, the "warhorse from Schaffhausen" was given an additional complication beyond the Grande Complication it already featured: a rattrapante or double chronograph. Only 125 of these precious pieces were made. In 1995, the Da Vinci also received its tenth watch hand, in the form of a rattrapante mechanism.

With the IWC Portuguese Rattrapante, IWC showcased a classic watch with an equally classic complication, worth considering in greater detail. This watch—which is also available in a three-hand version without complications and with chiming mechanism—owes its name to an order from company's Portuguese representative, who wanted this style of oversized watch for his home market in the late 1930s. The current models are historical replicas of this model, originally equipped with a pocket watch movement. Here, we are looking at another special feature of this watch. With a diameter of 44 mm, it is extremely large and, due to its proportions, only recommended for men's wear. However, women might want to wear this oversized watch as a special fashion statement.

The style of the case—bezel, push-pieces, and upper part of the lugs are polished, the rest is brushed—is strongly reminiscent of Lange watches, which, like IWC and Jaeger-Le Coultre then belonged, as Les Manufactures Horlogères SA (LMH), to the Mannesmann Group. The case back cover is secured with four screws and partly satin-finished and partly polished. The strap, with a width of 20 mm, is made from thoroughly dyed crocodile leather. It is attached with spring bars, as is the beautifully made clasp with the IWC logo. The dial is silver, the time display hands are gold-plated, and those for the chronograph are blued. The hour markers are applied as golden dots; the numerals, also in gold, are Arabic. The seconds display is in the form of an elevated bezel with black numerals and markers. The small seconds is at the six, the totalizer, which counts up to 30 minutes, is located at the twelve. The push-piece arrangement corresponds to the usual addition stopper, but a third push-piece at the ten shows that this watch can do even more. When the stop watch function is started and this push-piece activated, the second hand divides, as if by magic. This allows you to easily read intermediate times within a minute, while the actual time measurement is not interrupted. Another push of the push-piece, and the split-seconds hand jumps back over the second hand and again runs superimposed on it. This process can be repeated at any time.

The Big Pilot's Watch from 1940 with Caliber 52 T.S.C. was one of the largest IWC watches ever made. They even had an original pocket watch movement.

The IWC Museum displays the history of both product and company. It is houses in the landmark head office building. In the west wing, the first 100 years of IWC history are documented.

Otherwise, the chronograph works as usual. The movement, with IWC's own Caliber designation 35433, which makes this all possible, is based on the proven Valjoux Caliber 7760. The 7760 is the manual-winding version of the 7750, currently the most widely used automatic chronograph movement. IWC has, however, effectively re-engineered this movement. In addition to the gilding and various finishes, it shows, above all due to the design of the ratrappante mechanism, a traditional and thus high quality concept. While many manufacturers' rattrapante versions hide the single pliers mechanism under the dial, IWC offers a beautiful double pliers design on the back of the movement.

After the contract with F.A. Porsche Design expired, IWC introduced its own sports watch series with the GST line, which also featured such mechanical highlights as a rattrapante chronograph or a perpetual calendar, while bearing the same load as the normal watches this series, which only had a chronograph movement. The GST

Deep One appeared in this series in 1999; it has an integrated mechanical depth gauge. As early as 1994, IWC carried on the pilot's watch tradition with the Mark XII. Very similar to the legendary Mark II from 1948, now finally there was a typical IWC pilot's watch again. Four years later came the addition of the UTC pilot's watch with adjustable hours display and a 24-hour display. In 2000, when IWC, as well as Jaeger-LeCoultre and A. Lange & Söhne, was acquired by Richemont, they introduced, in keeping with the trend towards large watches, the Caliber 5000, which, with its Pellaton winding mechanism, reached a power reserve of seven days. It served as the drive for the Portuguese and the Big Pilot's watch.

With all the new products and all the optimism that prevails at IWC, the firm is nevertheless aware that the future needs a heritage, and that only those who can tell the consumer their story are successful in the mechanical timepiece market. IWC does not have to invent a history—it has one.

An almost sacral impression is achieved for presenting the artifacts by the lighting in the display cases. The two wing rooms that open to the left and right of the reception lobby radiate a cosmopolitan lounge atmosphere, well suited to the brand.

The archive and exemplary documents are on view in a small lounge area just behind the entrance and demonstrate the documented competence of the museum's exhibits.

Already by 1993, IWC, as part of activities marking its 125th anniversary, had set up its own small but exquisite museum on the top floor of the historical landmark parent company building. Special visitors were at times shown the as yet unseen select collectors' pieces. Nevertheless, the tradition of the brand and its products were still very much tucked away. That changed abruptly in 2007. IWC opened an entirely newly conceived museum on the premises of the former main building. There, IWC's product and company history are congenially and professionally presented on the ground floor in well-lit rooms.

High-quality metals, precious woods, white leather, and glass, plus refined lighting effects create an atmosphere that conveys the value expected by the IWC's customers and visitors.

The museum tells the history of the watch based exclusively on IWC watches. The illustrious pieces on view include a Jones Caliber from 1874, and a Pallweber Caliber with digital hour and minute display, both pocket watches, and one of the first IWC wristwatches from 1899. The proto-Portuguese from 1939 is to be seen, as well as the original pilot's watches. and the 1984 golden Portofino with moon phase display.

This clear cohesion of effect of yesterday, today, and tomorrow created by the museum's staging inspired the product designers, who introduced a Vintage Collection for IWC's 140th birthday. The platinum version, in an edition of 140 pieces, gives us, so to speak, the IWC Museum in a case to take home with you. This collection includes the manual-winding Pilot's Watch, with the same caliber as the Portuguese watch and the Portofino with moon phase, and the Aquatimer, Da Vinci, and Ingenieur, which are automatics. But back to the museum. In the west wing, visitors are brought closer to the first 100 years of the company and product history.

The walls are lined with display cases containing lovingly arranged artifacts. In the middle of the room is a central display case which allows you to access the phases of IWC company history by a thumbwheel. In the east wing opposite, interrupted by a central block where artifacts from the archive are presented, IWC watches made from 1967 on are presented against the background of their respective worlds of experience.

Yet the masterful fostering of tradition is, of course, more than just pretentious presentation to the outside world. Anyone who wants to present their history credibly and competently, must have an appropriate archive. Here again, they did a great job in Schaffhausen. A company archive and the scientific refurbishing of their own product and company history are part of the competent overall solution in terms of tradition. The history of IWC calibers and watches is meticulously documented and made available electronically.

In the new tonneau-shaped Da Vinci, with its refined chronograph mechanism, IWC presented in 2007 a completely newly developed in-house-crafted caliber—a column wheel chronograph with automatic winding. As the special Edition Kurt Klaus series, it was also made with perpetual calendar as a tribute to long-time IWC chief designer and the spiritual father of the Da Vinci.

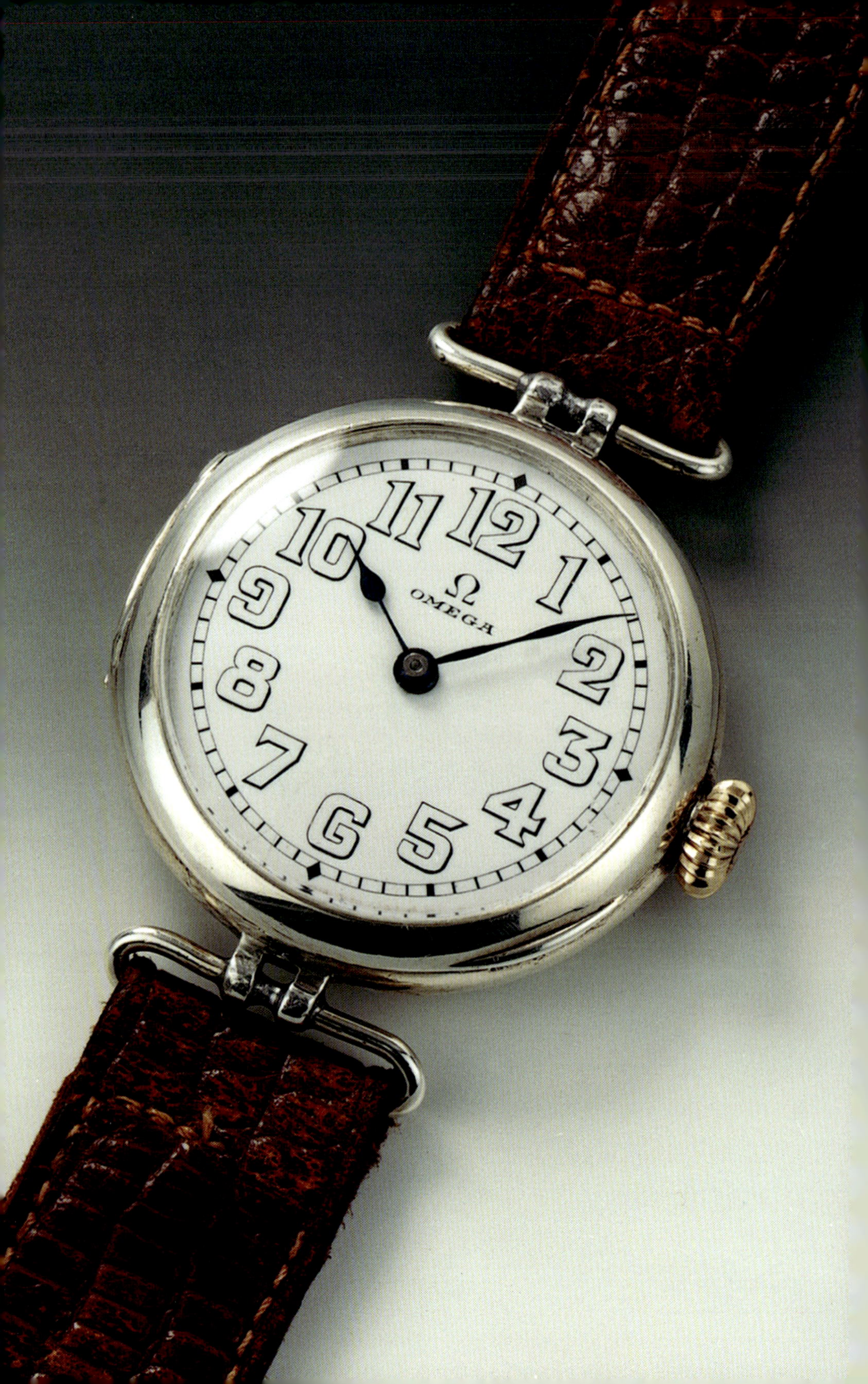
OMEGA

From Alpha to Omega—The Omega Museum

Omega is one of the core brands of the Swatch Group, a company that was for decades the flagship of the Swiss watch industry. The Omega name meant precise chronometers that brought home more Observatory prizes than many of its competitors. Since the beginning of the 20th century, sports timekeeping has been the basis of the company's popularity. Whether at the Gordon Bennett Balloon Race in Zurich in 1906 or at the Olympic Games (their first was in Los Angeles in 1932)—Omega has been the timekeeper.

The Omega brand name was created based on the idea of finding a distinctive name which symbolizes the purpose of the product. Omega, the last letter of the Greek alphabet, has always stood for perfection. This was exactly the message they sought to communicate. Louis Brandt, founder of the company, would not live to see it named. Since 1848, he had been assembling watches from purchased movements under his own name in La Chaux-de-Fonds (Neuchâtel).

After his death in 1879, his two sons, Louis-Paul and Caesar Brandt, moved the company to Biehl and there developed it into a watchmaker that produced its own movements. The brothers also established the brand name Omega in 1894. Previously, the watches had carried product names such as Helvetia, Jura, Patria, Celtic, Gurzelen, or Labrador. This latter watch was equipped with the company's first mass-produced caliber, which brought it to an accuracy of 30 seconds per day. In 1898, the company Louis Brandt & Frères, with about 500 employees, was making 100,000 watches a year.

A typical silver Omega wristwatch from the 1920s; the lady in the poster is still wearing her watch on a chain around her neck.

Relatively early on, Nicolas G. Hayek recognized the importance of tradition for the Omega brand name, which had, by the end of the 1970s, seen its mystique, developed over almost 100 years, be run into the ground due to inflationary production and distribution policies. As one way to counter this, the Omega Museum was opened in December 1983. The museum presents exhibits of watches, watch movements, table clocks, instruments, tools, photos, engravings, and awards, from the abundance of the Omega collection of some 4,000 pieces.

This includes the founder Brandt's own worktable and lamp. The focus of course is on presenting historical Omega watches. In addition to pocket watches, such as the "pastime" watches popular in the 19th century, with devices to set roulette or horse races in motion, there are extravagant pieces

Precise chronometers were Omega's hallmark already in the pocket watch era. Devilishly good and accurate, as the promotional poster suggests (right).

HORLOGERIE
DE
PRÉCISION
SPÉCIALITÉ :
MONTRE „OMEGA"
OMEGA
E. DONAUER
BAHNHOFSTR. 48 ZÜRICH.
GRAPH. WERKSTÄTTEN GEBR. FRETZ A.G. ZÜRICH

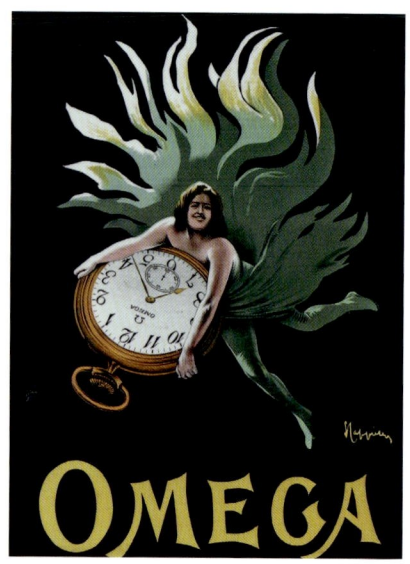

Before the Second World War, Omega watches had a status and prestige matching that of Rolex today. Anyone wanting a robust and accurate sports watch would go for Omega.

to be seen. One is called the Greek Temple, a richly ornamented gold pocket watch; another is one of the first mass-produced wristwatches in the world, which was awarded the Grand Prix at the World Exhibition in Paris in 1900. There is a martial presentation of various military watches. In addition, there are such interesting pieces as the pocket-watch-size chronograph, which looks modern even today, worn by Lawrence of Arabia. Of course, the theme of the Olympic Games and Omega's role in sports timekeeping is thoroughly discussed, and the pocket chronograph which launched timekeeping at the Olympics is on view, as well as the first photo finish camera from 1949. The Omega Marine is one of the first diver's watches in the world. Commander of Apollo 11 Neil Armstrong's "small step for man and giant leap for humanity," on July 19, 1969, was done with an Omega Speedmaster on his arm. As a result, the company can advertise with confidence that the Speedmaster is the only watch ever worn on the Moon. Since 1963, this model has been the official timepiece of American astronauts, and, since 1975, also of the Russians. The selection of this watch was made rather by chance. NASA sent an employee to some watch shops in Houston, who brought back all sorts of manual-winding chronographs with him. These were first subjected to an elaborate series of tests, and, by the end, the Speedmaster Professional was the clear winner. The space suit used from 1969 to 1972 by American astronauts is on view, as well as a command console from the big control room of the Houston Space Center

The military consistently played a role in the general acceptance of the wristwatch. Here are two versions. The one on the right has a metal grid to protect the crystal.

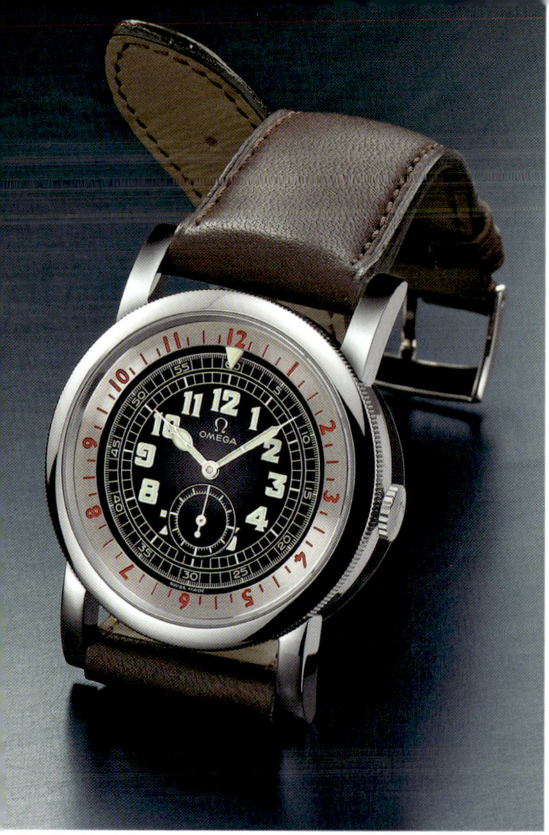
OMEGA

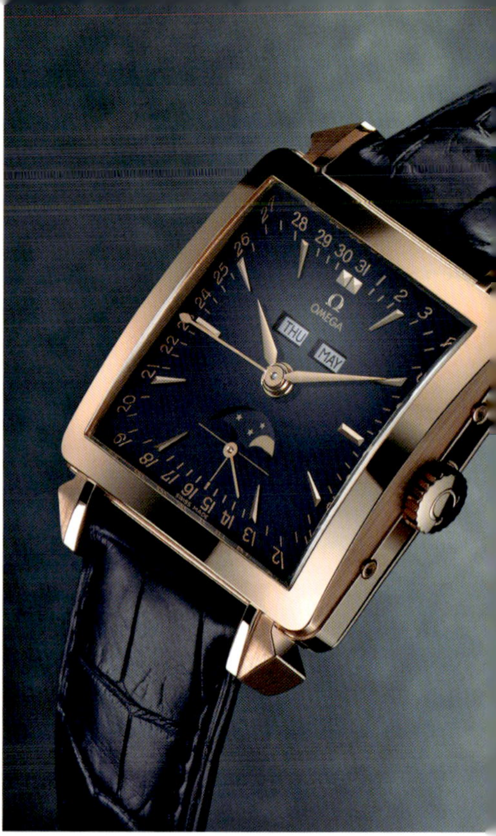
OMEGA
THU
MAY

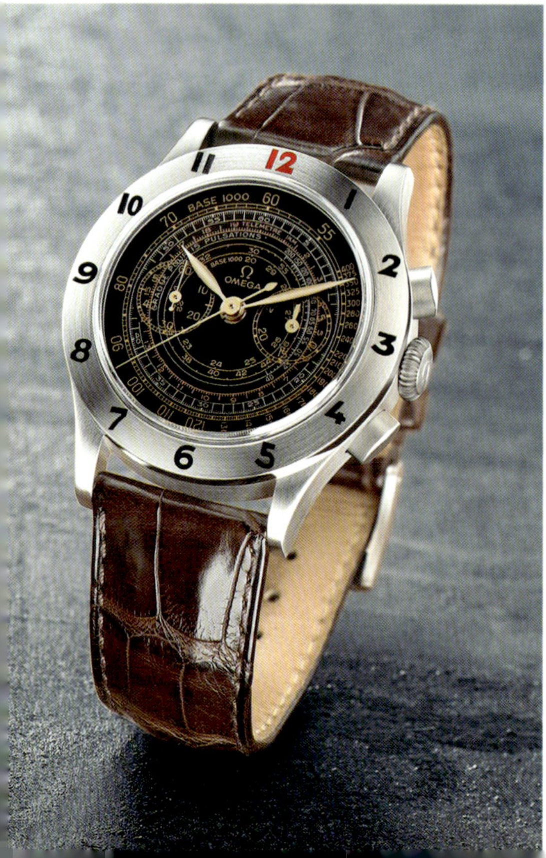
BASE 1000
PULSATIONS
OMEGA

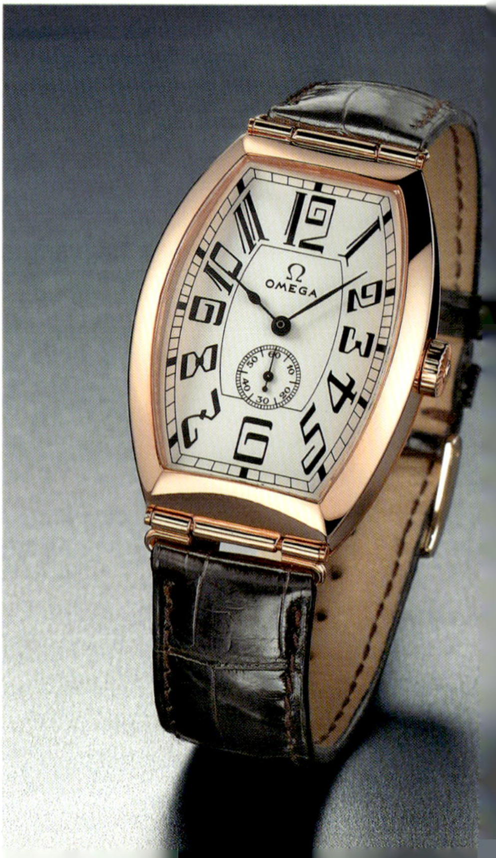
OMEGA

from the time of the Mercury, Gemini, and Apollo missions. In addition, the interested public can see high-tech quartz innovations, such as the Omega Marine Chronometer from 1974, the most accurate quartz wristwatch ever made.

The most important contemporary model lines offered by Omega are the Constellation, the Speedmaster (first watch worn on the Moon), the Seamaster, and the De Ville as an elegant addition to the range. These series include both automatic and manual-winding watches, as well as watches with quartz movements,

Omega's orientation to successful past models is demonstrated by such watches as the De Ville series manual-winding chronograph. This watch, like its predecessors, is based on a minimalist and traditional concept. The shape is reminiscent of the 1950s, and one is tempted to speak of retro style. At that time, Omega had a similar model in its production program, the Seamaster. The De Ville series was also already available in the 1970s—even featuring the same movement, although without an hour totalizer. The new version has Dauphine hands, and

Omega watches from the Museum Edition, which are modeled after historical models: Pilot's watch, chronograph, calendar watch, or an elegant tonneau watch in rose gold, present classic design with modern technology.

even the sub-dials have applied gold baton markers; the dial is silver plated with *Rayons de Soleil* (sun ray) design. The totalizers for the chronograph functions are at the three and six; they count hours and minutes. The small seconds is at the nine.

This arrangement, in combination with the manual winding, already indicates the movement, which also comes from a Swatch Group movement factory. It is a Lemania 1873 Caliber with 18 jewels, 21,600 vph and a power reserve of 45 hours. The movement has a diameter of 27.5 millimeters (equivalent to 12 lignes) and a height of 6.9 millimeters. It is equipped with a Nivarox flat balance spring with fine adjustment. The board is perlaged; the bridges decorated with *Côtes de Genève* or Geneva stripe. It is a shame that this watch does not have a transparent display back, since this movement is worthy of daily contemplation.

Omega re-engineered this movement, also used in the Speedmaster Professional, and awarded it its own Caliber designation: Omega 861 This is common practice today, even among those manufacturers who demand ten times the price. Those mechanics aficionados in search of this chronograph movement with cam-switching, will get their money's worth with the De Ville, in terms of price-value ratio. This watch is available, like all Omegas, only in pure steel or gold or a combination of the two. Gilding has been passé at Omega since 1989.

Its wearability is also enhanced by the medium-sized case with 36.1 millimeter diameter and 11.4 millimeter height, and the low overall weight of 55 grams. Besides the bezel, crown and push-pieces are made of gold. The push-pieces have a clearly identifiable pressure point and require a certain amount of pressure. But they are easy to use, which is especially important in a manual-winding watch. Here the look and feel are essential—the De Ville gives the sense of wearing a solid timepiece on your arm. Dial, hands, and movement are precisely fitted. The timekeeping rate results of this Omega model are very good, at plus 4 seconds per day; however, as with most manual watches, this depends on regular winding,

The case back cover is only snapped on, not screwed down. Nevertheless, the watch is waterproof to three atmospheres, so it would be quite suitable for swimming. The case back cover has a brushed finish and bears the watch's serial number and trademark symbol.

Anyone looking for a chronograph without the martial stance of a Breitling or Rolex, will find the Omega De Ville an interesting alternative. A watch that fits well under a shirt cuff and does not look bulky.

Omega has lots of technological advances to offer, such as the Co-Axial escapement developed by George Daniels. This escapement combines the advantages of a lever escapement with those of a chronometer escapement, including essentially no need for lubrication. Omega installed this escapement in both its chronometers and chronographs, and again set a trend with it.

As a result, meanwhile, we now have the De Ville chronometer also with automatic movement and Co-Axial escapement. The Museum also exhibits this revolution from 1999, making a visit a must for any watch enthusiast and especially fans of the brand.

"He, too ...". John F. Kennedy wore this rectangular Omega during his term as American President.

PRESIDENT OF THE UNITED STATES JOHN F. KENNEDY FROM HIS FRIEND GRANT

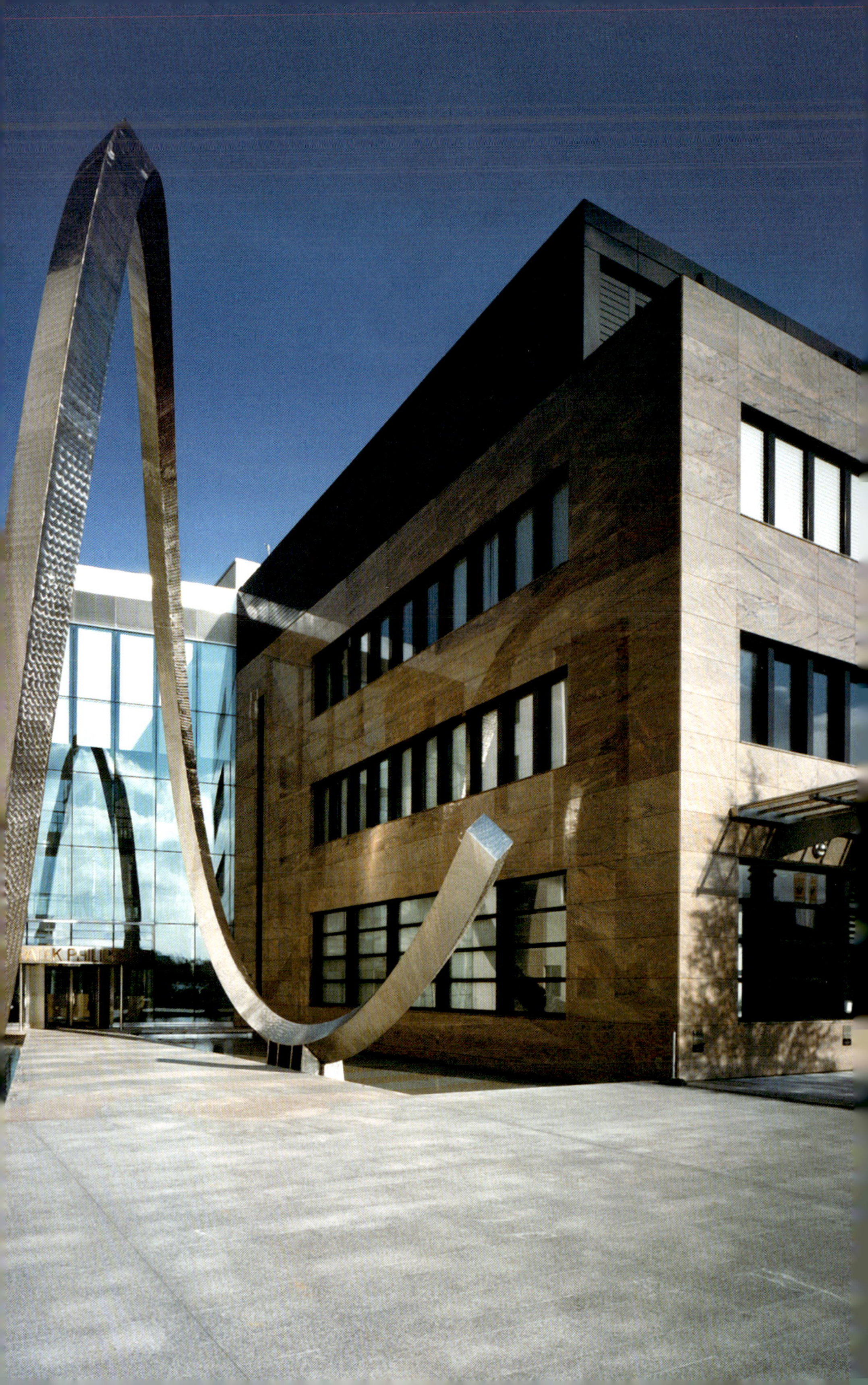

Brands

There's Only One at the Top—Patek Philippe, the Watchmaker

No other make of watch achieves any higher price at auction for historical models as Patek Philippe: over $4 million for a Reference 1415, a world time watch from 1946 in a platinum case. No company exudes such an aura, created by both expert craftsmanship and history. In the highly competitive market for mechanical timepieces, Patek Philippe is the absolute exception. Seemingly unaffected by any economic cycle, these watches are allocated rather than sold. There are some manufacturers, who dub themselves the "master of complications," but in truth, there are few who can compare themselves to any extent with the Geneva watchmaker. The quest for quality, in combination with an extreme vertical range of manufacture, gives these watches their unique flair.

But this was not always so. Even Patek Philippe was affected by the quartz shock. Then-head Henri Stern believed at the time in holding fast to the electronic watch. For him, precision was the primary seal of quality for a high-priced watch, and traditional mechanical watches could not hold a candle to their quartz counterparts in this respect. Just like Rolex, the company began to produce their own high quality quartz movements in 1956 and put them in men's and women's watches. Yet Patek Philippe did not neglect mechanical watches, which, as things developed, turned out to be a blessing.

An oversized stylized coiled spring in front of Patek Philippe's new factory building symbolizes both its in-house manufacturing concept and commitment to the mechanical watch.

As the mechanical boom began in the mid-1980s, the Geneva-based company became an icon of mechanical watchmaking. The Caliber 89, commissioned on the occasion of the brand's 150th anniversary, did the rest to reinforce this reputation. This watch displays the starry heavens over Geneva on its case back, along with sun hand, equation of time, sunrise and sunset, the date of Easter, and the Zodiac. Seen from the front, the watch informs its owner of the date by retrograde display, and of course the perpetual calendar shows year and leap year. The 24 hands and 12 sub-dials provide more information. A rattrapante chronograph, temperature display, alarm, and autonomous and settable chiming mechanism, complete this mechanical marvel. It includes 1,728 parts and has 33 complications, and with it, Patek Philippe set new standards for watchmaking. That this watch, made as a pocket watch, was also an acknowledgement of traditional watchmaking value. It was a successful appreciation of the company's anniversary as well as good PR.

Everything began on May 1, 1839, in Geneva, when watchmaker François Czapek and businessman Count Antoine Norbert de Patek, both Polish immigrants, founded the company Patek, Czapek & Co. After a short time, Patek separated from his partner and entered into a new relationship with French watchmaker Jean-Adrien Philippe, who had introduced the crown winding mechanism to watchmaking. Until then, a key was required to wind the watch. The first watch, which used the crown both for

The company name Patek Philippe & Co first appeared on Jan. 1, 1851. The first company building on the shore of Lake Geneva.

winding and setting the hands, was made in 1844. A year later, the company was renamed Patek & Co., and made the first watch with minute repeater function that same year. In 1851, came another company name change, one that would endure to this day: Patek Philippe & Co.

From its beginnings, the Geneva company established complicated mechanical watches as its domain. In 1902, the firm presented a double chronograph; in 1909, the Duc de Regia, a watch with small and large Westminster chimes; and in 1916, it even produced a petite ladies' watch with five-minute repeater. As early as 1868, Patek Philippe had made a wristwatch for the Hungarian Countess Koscowicz, and can therefore claim to have made the first Swiss wristwatch. Starting in 1927, Patek Philippe also offered, in a wristwatch, the complication that is a distinctive feature of a century fascinated with speed: the chronograph. Already two years earlier, they had manufactured the world's first wristwatch with perpetual calendar. However, despite

Next to the Museum, the old head office still serves to represent the firm, and is well worth a visit as you follow the manufacturer's footsteps through Geneva.

Reference 5002: the Sky Moon Tourbillon in gold and platinum—one of the most complicated wristwatches in the world. In addition to complications such as perpetual calendar, minute repeater, tourbillon, and moon phase, on the back sidereal time, a star map, and the phases and angular movement of the moon can be read.

all their innovations and excellent products, this traditional company did not remain unscathed by the global economic crisis. It required new capital.

Since 1901, the Ancienne Manufacture d'Horlogerie Patek Philippe & Cie. S.A. had been a corporation. The brothers Charles and Jean Stern, previously the dial manufacturers for Patek Philippe, took over the majority of shares in 1931. Now in its fourth generation, the Stern family has been directing the company skillfully and wisely through all the economic cycles and passing fashions. In 1932, under the aegis of this family, the Calatrava model series was introduced. The Calatrava Cross, emblem of Abbot Raimondo, who, in 1158, founded an order of knights in the city of Calatrava, has even become a symbol of the firm. Like Rolex, Patel Phiippe is committed to continuity for its model families, which are manufactured for several decades. This includes the Golden Ellipse, first introduced in 1968, with blued gold dial and elliptical case, which became a distinctive emblem of the fashion jet set. In 1976, with the Nautilus, a sports watch designed by Gérald Genta with a bezel modeled on a porthole, the company presented a steel watch, waterproof and suitable for sports. Waterproof down to a depth of 120 meters, this was a watch for everyday wear that could be worn on any occasion. And that is how it was advertised: "They go as well with a wetsuit as they do with a dinner suit." The advertising flirted with the high price: "One of the most expensive watches in the world is made of steel," as one ad headline read. The Reference 5980/1A, at 44 mm a watch with unusually large proportions for Patek Philippe, is today one of the Nautilus family. The heart of the watch is the company's first in-house chronograph movement, which can be admired through a sapphire crystal case back. This column-wheel controlled chronograph caliber has a flyback function and a power reserve of 55 hours. The chronograph display of minutes and hours are combined in a totalizer.

In 1993, the new Gondolo collection was introduced, which features Art Deco style in a rectangular or tonneau shape. The various movements range from a three-hand watch to a perpetual calendar. In 2000, the firm carried on the tradition of the highly complicated pocket watch with the Star Caliber 2000. This watch, like the Caliber 89, is a double-sided pocket watch with 21 complications.

A year later, Patek Philippe offered a comparable model for wrist wear with the Sky Moon Tourbillon. This is again a double-sided watch, the first by Patek Phillippe for the arm, which shows on its front the mean solar time, an eternal calendar with leap year indicator, and the age of the moon. If you take it off your wrist and turn it over, the Reference 5002 shows the northern nocturnal sky, phases and angular motion of the moon, and sidereal time. The chimes and tourbillon go a step further to enhance the complexity of this miniature marvel. As indeed all mechanical movements by Patek Philippe do, this Reference also features the *Poinçon de Genève* (Geneva Seal), which is only awarded to watches made to the very highest standards.

While at Patek Philippe there is a strong sense of being bound up with the mechanical watch and the associated traditional manufacturing methods, at the same time, they are seeking new pathways. In 2005, the company's Advanced Research team presented for the first time a watch with low friction silicon escapement wheel.

Amidst the large-scale consolidation process in the watch industry, Patek Philippe remains one of the few independent watch manufacturers. It is the Stern family's stated goal, that this is how things will remain. Anyone keen on Patek Philippe is in good company. Queen Victoria wore one of their watches, as did Richard Wagner, Peter Ilyich Tchaikovsky, Rudyard Kipling, Leo Tolstoy, as well as the man who revealed the relativity of time: Albert Einstein.

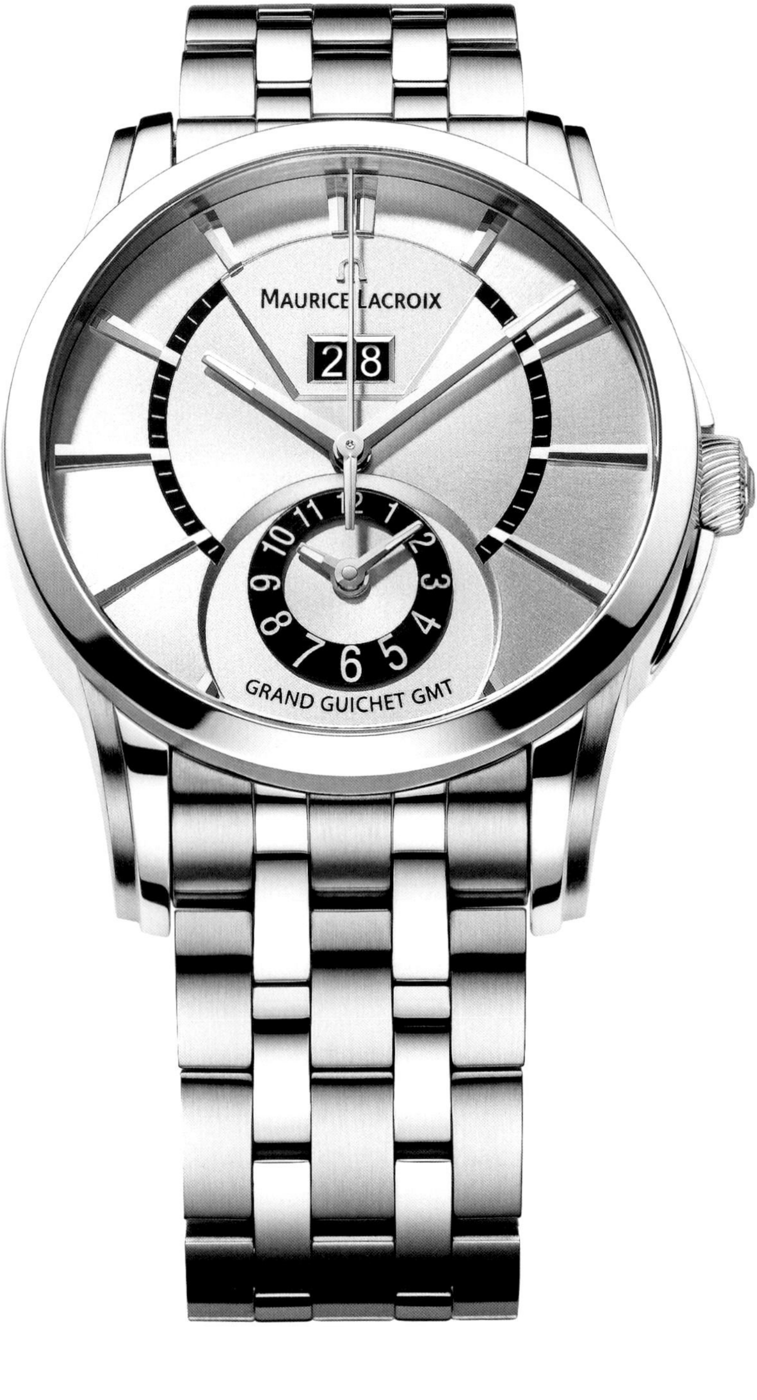
MAURICE LACROIX
28
GRAND GUICHET GMT

Maurice Lacroix—The Way to In-House Watchmaking

Amidst all the brands with their 100 years of history, it isn't easy for a newcomer. Nevertheless, the young brand Maurice Lacroix has managed to position itself alongside the established brands today. Once it had emerged from the over-100-year-old Zurich trading house of Desco von Schulthess, when an assembly operation was set up in 1961, the success story began for Maurice Lacroix, which was first presented in 1975. The product range extended from quartz watches to mechanical special features, in an extraordinary price-performance ratio. Watches such as a wristwatch alarm with basic Caliber AS 5008 or chronographs with the Venus 188 movement, are now considered collectors' items.

The brand reached another milestone in 1989, with the acquisition of the Queloz watch case factory in Saignelegier, where highly complicated watch cases are still being made today.

By 1993, the steady growth of the German market made it necessary for the Pforzheim-based German subsidiary to move into an imposing office building. By offering high-quality mechanical complications at affordable prices, the brand contributed significantly to the renaissance of mechanical watches. The highlight of the former Les Mécaniques series was the manually wound chronograph in gold with an old Valjoux 7736 movement. For several years already, Maurice Lacroix had been focusing on developing attractive and innovative additional functions for their exclusive mechanical masterpieces. This laid the conceptual foundation for its future top-of-the-line collection, which soon was named the Masterpiece Collection.

With the new Caliber ML 106, Maurice Lacroix went one step further. Until then, high-quality movements from earlier eras had been re-engineered and fit into appropriate cases; now, with this new design of a classic mechanical manually wound chronograph, the firm took the step to *Manufaktur*, or in-house fine workshop watchmaking. The traditional theme of the chronograph, with technical parameters set since 1862, was newly interpreted by Maurice Lacroix designers, taking into account both modern technology and contemporary modifications. Most conspicuous is the immense size of the movement. A glance through the sapphire crystal case back shows a 36.6 mm diameter movement in its full glory. An automatic winding mechanism was consciously omitted, since its oscillating weight would have greatly

Maurice Lacroix's interpretation of a GMT watch with additional big date.

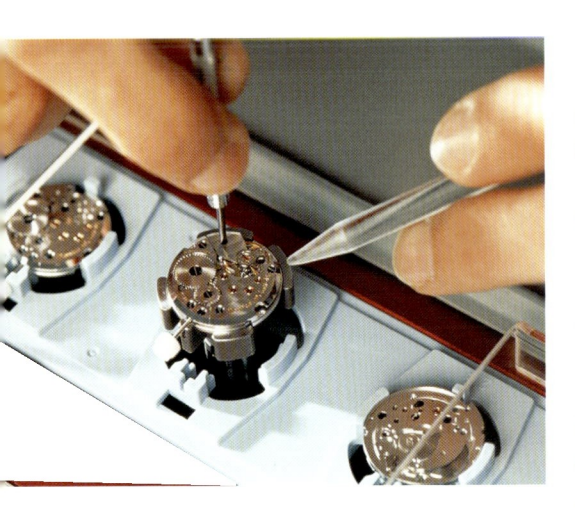

"More in-house movements and creation of value" is the motto of this fledgling watch company.

obscured the view of most of the interplay of all the wheels and levers.

The watch has a 48-hour power reserve, and daily winding should be a pleasurable moment amidst everyday hustle and bustle. Actuation of the start push-piece creates a connection between this constantly ticking microcosm and its additional mechanisms— "motor" and "transmission" are linked. This connection is made, as in a car, by a coupling, or clutch. The horizontal coupling is regarded as state-of-the-art.

By starting the chronograph, a finely toothed wheel pivots between the watch movement and stopwatch. Due to the balance wheel frequency of 2.5 hertz (18,000 vph), the large second hand goes into motion in fifth-of-a-second steps, readable by the 300 subdivisions on the dial. Analogously, removing this connective bridge causes the hand to stop instantaneously. The chronograph center wheel with its teeth halts, while the movement keeps running.

Maurice Lacroix created an exemplary solution for the control of these and other functions in the ML 106 Caliber, namely, by column wheel. Another special feature of this watch is the interpretation of the minute totalizer. The vast majority of chronographs with no hour markers use totalizers, which count up to 30 or 45 minutes maximum. In the Masterpiece Le Chronographe, Maurice Lacroix extends this time period to an unusual 60 minutes. This rare 60-minute counter required a significantly greater horological investment.

The traditional chronograph movement with column wheel and swan-neck fine adjustment is among the best that the market has to offer. This precious piece is set in a red gold case. As with the Masterpiece Vénus, launched in 2004, all operative chronograph parts are polished and adjusted by hand at Maurice Lacroix. With the ML 106, we can say with confidence that the Maurice Lacroix company has raised itself to the peerage of in-house watchmaking.

A calendar watch with a difference. Date and weekday display by central pointer.

UNITS PER HOUR
ROLEX
OYSTER PERPETUAL
SUPERLATIVE CHRONOMETER
OFFICIALLY CERTIFIED
COSMOGRAPH
DAYTONA
SWISS MADE

Every Movement Gets a Crown—Rolex, the World's Largest Luxury Watchmaker

The French sociologist Claude Levi-Strauss coined the phrase: "Myths are hollow and round." Anyone who breaks the outer shell ultimately is left holding nothing in their hands. Still better perhaps is the image of an onion, of which nothing remains after the last layer is peeled away. Myths about companies and their products are also made up of layers and shells from history and stories, both consciously staged and fortuitous, which achieve meaning only in retrospect. The mythos of a brand and its products only unfolds in concentration, the meaningful interconnection and knowledge of the distinctive quality of events. Especially in this era of ever-more extensive industrial production, it is the trademark that makes it possible for the consumer to orient themselves and distinguish products.

Among those iconic brands which started on their road to success at the beginning of the 20th century, is undoubtedly an art term registered as a brand name in 1908, Rolex. This name was invented by its German company founder Hans Wilsdorf (1881–1960) from Kulmbach; it is said to derive from *"horlogerie exquise,"* although this cannot be proven. The company founder made it his goal to present a precision wristwatch that could compete with the pocket watches then in general use. It was also he who defined the basic values of the brand, which define its essence to this day: the highest quality craftsmanship in manufacturing and processing, optimum precision, while also robust and suitable for everyday use, including models—designed for different applications—able to withstand extreme wear. In the Rolex mythos, Wilsdorf played a similar role as Enzo Ferrari did for the eponymous car brand. One wanted to present the world with the best sports and racing cars; the other the most accurate and sturdy watch. But Wilsdorf was more than watch technician and maker. He was also a gifted press and marketing specialist, a craft, which at that time operated under the label "propaganda." He quickly realized how to use people as testimonials; people who conveyed the message of his products. With

Caliber 4130, encased in the new red gold alloy Everrose, also chlorine resistant. The Daytona is still one of the most sought-after Rolex models.

tremendous willpower and determination, provided with a sure instinct, he put his ideas and ideals into practice.

Yet his biography began with a severe stroke of fate. At twelve, he lost both parents in quick succession, first his mother, and then his father. Relatives now took care of him and his two siblings; they sold the family business, profitably investing the income received. These were relatives from the mother's family, the Bavarian brewing dynasty Meisel. What would be more obvious, than to make little Hans into a capable brewmaster? But there was nothing that Wilsdorf wanted less. After a period at boarding school in Coburg and finishing his *Abitur* (graduation examination), he completed a business apprenticeship with a man in Bayreuth, who operated a global trade in artificial pearls. Business suited him, and at the same time he had great interest in both watch technology and foreign languages. So, at almost 20 years old, he went to Switzerland, to a large watch exporting firm in La Chaux-de-Fonds. For 80 francs a month, he undertook the English correspondence, carried out office work and also wound up the pocket watches in which the company dealt. Precision timekeeping was fast becoming his obsession. With a portion of his paternal inheritance, he bought three gold pocket watches and had their accuracy recorded at an observatory with precision certificates. Then he sold the watches at a profit. Here already Wilsdorf's business principles were becoming apparent. But before he moved to London, which was the center of the industrial world in 1903, Wilsdorf finished his one-year military service in the German Imperial Army. In 1905, after he had accumulated sufficient experience in the watch trade, he founded, together with the much older Alfred James Davis, his own watch dealership company under the name Wilsdorf & Davis in London. He got some of the capital from his brother and sister, but during the crossing to England, 30,000 gold marks of his patrimony was stolen.

While pocket watches had aroused his interest and enthusiasm for the world of watches, it was now the wristwatch, the innovation of the industry, which was becoming a developing market. At Aegler in Biel, he bought so many high-quality lever movements that the sum to be paid amounted to five times the firm's capital. Yet his success proved him right. As early as 1907, the company opened a subsidiary in La Chaux-de-Fonds. In 1908, Wilsdorf & Davis was among the largest companies in the watch trade and had 200 models in its product

A Rolex Precision with full calendar and moon phase. A complication no longer manufactured by Rolex. The Reference 8171 is also much sought in the collectors' scene.

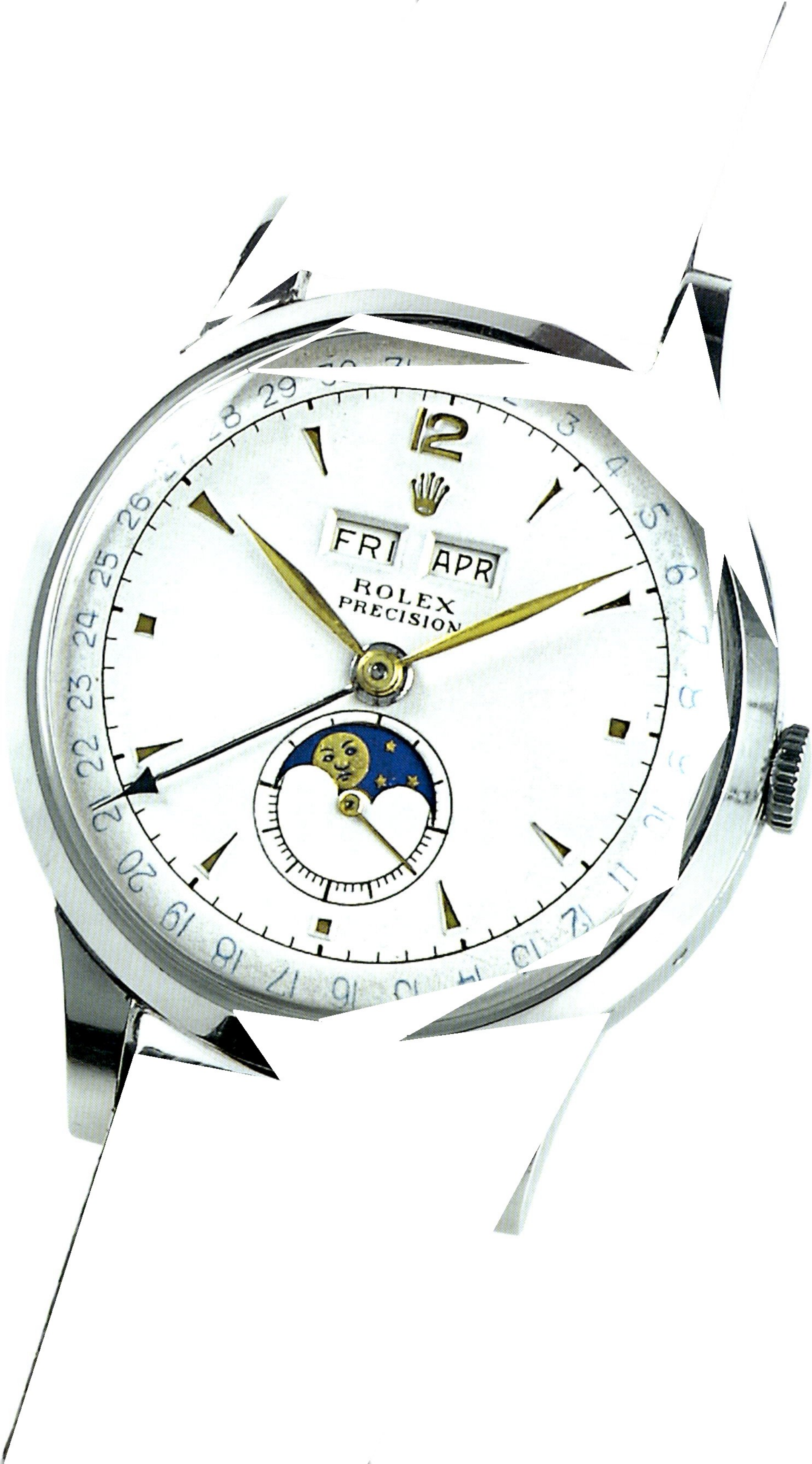
FRI
APR
ROLEX
PRECISION

range. The watches were sent for sale anonymously, or with the logo of the respective distributor. Only the cases were stamped with W/D, for Wilsdorf & Davis. This displeased the patrons, as did the fact that wristwatches at this time were still considered to be ladies wear, and entirely unmanly.

As a first step, he thought up a product name for himself, about which he writes in his reference book: "It was so short and yet so catchy, that there was even enough space for the name of the English dealer next to it on the dial. But what was particularly valuable is that ROLEX sounds good, is easy to

Reference 6239/6263 with black dial and red Daytona inscription and tachymeter scale, was Rolex's last manually winding chronograph; it was manufactured until 1985.

remember and is also pronounced the same in all European languages."

It would be another 20 years until the new name had finally established itself. Wilsdorf initially used a trick. In the cartons of six pieces, only two watches were labeled Rolex; later, it became three, and so the name could become known in the dealers' show windows. Now it was time for the young company to furnish proof of the quality of small movements. Could the dainty ladies watches stand up to the respectable chronometers in men's pockets in terms of accuracy? They could! Already in 1910, Wilsdorf had received a first class certificate in Biel for a wristwatch with a 24.81 mm movement diameter. The watches repeated the feat in 1914 at the Kew Observatory in England, to obtain a Class A certificate of accuracy. This meant that the "little watches"—considered such by contemporaries and met with hostility by watchmakers—operated with the timekeeping performance of a marine chronometer. Suddenly, Wilsdorf joined the circle of England's most prestigious watchmakers—and he achieved this with a wristwatch.

From the beginning, Rolex's blank movement maker was Aegler of Biel, behind which stood Jan Aegler with his company founded in 1878. Since 1881, the company has had its headquarters at Rebberg, near Biel, and since 1900 had been exporting ladies wristwatches all over the world, except, beginning in 1913, to England, so as not to compete with its business partner. In 1914, the company was incorporated under the name Aegler SA, Rolex Watch Company. It employed 200 staff and is the exclusive supplier for the Wilsdorf & Davis Rolex Watch Company. As a company, Aegler remained independent, a condition which only changed in 2004, when Rolex President Patric Heiniger bought it for the princely sum of 2.5 billion euros and integrated it into Montre Rolex SA.

"Bellum omnium rerum pater est" ("War is the father of all things") said the Romans, and this is also true of wristwatches. Both the frequent British colonial wars and the First World War shifted watches to the wearer's wrist, where it could be read quickly and easily. This development brought the Wilsdorf & Davis Rolex Watch Company its wished-for business success. The increase of import tariffs to 33.3 percent in England in 1919, brought the company to an end. Export operations were transferred to the Biel office, and Wilsdorf himself moved to Geneva with his wife. In 1920, the firm was renamed Montre Rolex SA, after Wilsdorf split from his disliked partner. From then on, movement manufacturing continued in Biel, while case production and assembly was done in Geneva. In 1925, a trademark symbol, in the form of a five-pointed crown, was added to the brand name; it has adorned all Rolex watches from 1939 until today.

The company's history is marked by a tenacious struggle to perfect the wristwatch, always confirmed anew by official certificates. The observatories of Kew, Geneva, and Besançon became the places to go to certify the accuracy of small movements. The

Observatory in Neuchâtel even distinguished a women's chronometer with a movement diameter of 15.23 mm with a first class certificate. Rolex has continued its policy of making observatory-tested watches up to the present. Rolex remains the watch manufacturer with the most certified chronometers in the world. Already by 1968, Rolex had reached the one million mark for chronometers; today, there should be a total of over 20 million Rolex-made watches with chronometer certificates.

For mechanical watchmaking, water has always been a natural enemy of this mechanical marvel, initially made of iron. It was not until the 1920s that a case design appeared on the market that made it possible to hermetically shield the movement against external effects. Montre Rolex SA was a trailblazer of this development. In addition to mechanical robustness excellent precision, it was Wilsdor's especial declared goal to make a watch that would be absolutely waterproof. This was achieved through sealed case parts screwed down against each other, a special screw-down crown design, and an interlocking crystal. The name for this work of art was quickly found: Oyster—the oyster as a symbol of the hermetic seal. By 1926, patent applications were made in the England and Switzerland. The patent for the first Oyster was issued in 1926, and this innovation was introduced to the astonished public.

When the young English typist Mercedes Gleitze swam the English Channel on Oct. 27, 1927 in 15.5 hours, Wilsdorf took notice of the young, extreme sportswoman. When she made a second swim at end of October, she was wearing an octagonal Rolex Oyster on her wrist and thus made it clear to all the world that the breakthrough of a waterproof watch had been achieved. Although Mercedes, due to extreme cold, had to give up after about ten hours 11 km off the coast, it is a success that a watch had withstood the elements for so long. Wilsdorf announced this triumph in a full-page ad on the front page of the *Daily Mail* that cost him 40,000 Swiss francs. Rolex was suddenly all the rage. As another publicity stunt, Rolex used small aquariums, in which concessionaires could display the watch with goldfish swimming around it, to astonished window shoppers. Very quickly, however, it turned out that the screw-down crown was the weak point of the design, since it had to be opened and closed for daily winding the watch, subjecting it to considerable wear. Therefore,

The Rolex Prince with shaped movement has been made in various versions since the 1930s. Here, an aerodynamic chronometer from 1950. Reference 3361 enjoys great popularity among collectors.

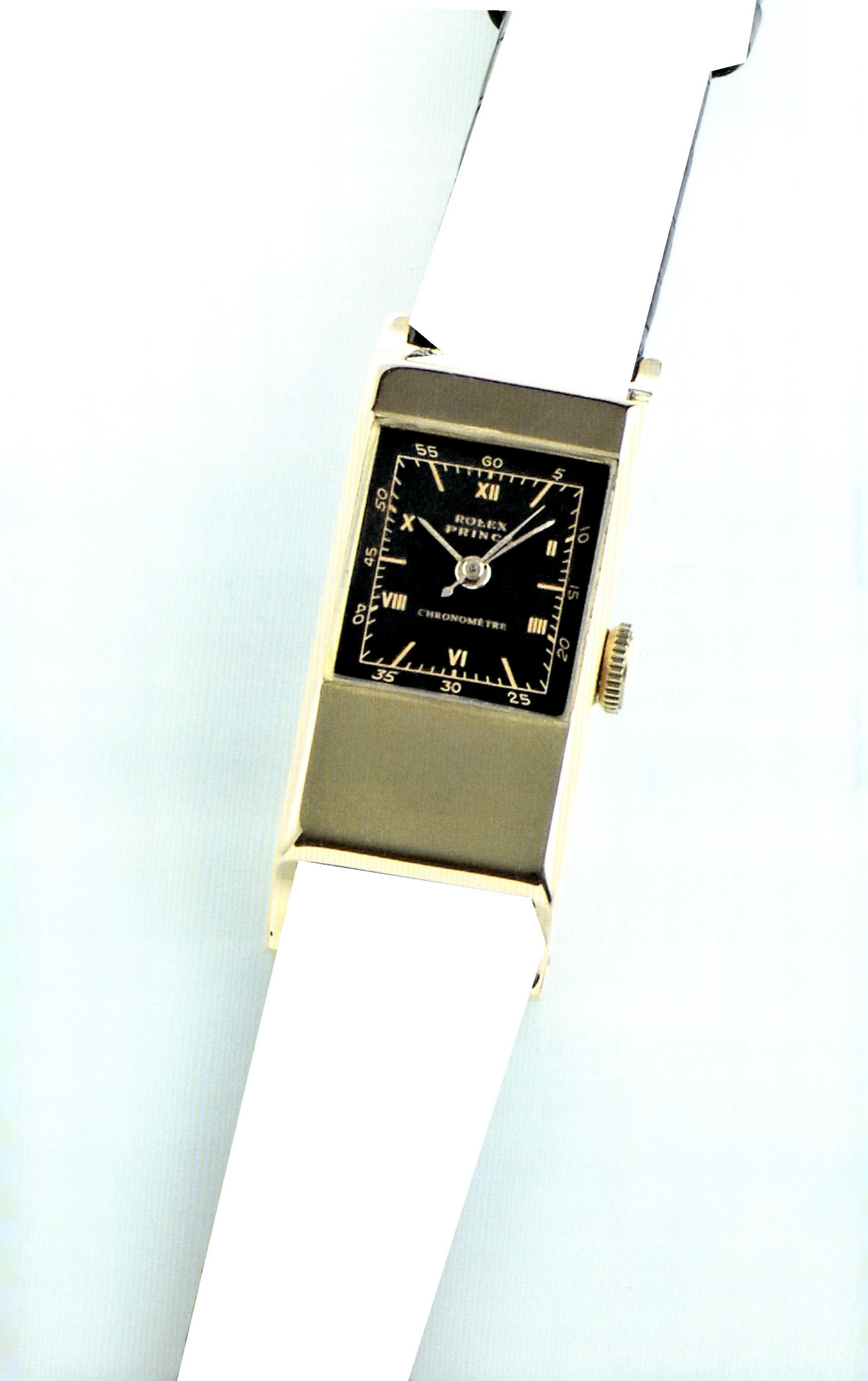
ROLEX
PRINCE
CHRONOMÈTRE

in 1931, Rolex developed an automatic winding mechanism, patented in 1933. The Ref. 1858 is the first Rolex with rotor mechanism. After six hours of wearing time, the watch is fully wound, according to the proud advertisement, which also notes that the watch can also get its energy supply, as before, via the crown. In 1945, a date mechanism which jumped exactly at midnight went into series production with the birth of the Datejust; the same year, Rolex's 50,000th chronometer was certified in Biel.

The chronology of achievements can still be found today on every Rolex dial. "Superlative chronometer officially certified" recalls the strict testing criteria; "Oyster" the waterproof case construction; "Perpetual," the Rolex-developed automatic winding mechanism; and "Datejust" the date which jumps exactly at 00 o'clock: the compressed

Sports watches in gold were the Rolex hallmark for many years. Here, a GMT, also available in a less pretentious steel version and which enjoyed great popularity among pilots.

history of a long technical development process is documented in the smallest space. But Rolex also makes luxury watches, less intended for sportswear. The Rolex Prince, a rectangular watch with a shaped movement and separate second hand, which came on the market at the end of 1928, aimed at the world of the rich and famous. "The Watch for Men of Distinction," according to contemporary advertising. For King George V's Silver Jubilee, 400 of these watches were ordered—as certified chronometers, of course. This model has been so successful that it has remained in the production program for 40 years—it is also a hallmark of the company, that it does not unsettle its customers with hectic model changes. Rolex's internal record is held by the Submariner, which has been manufactured for 50 years, without a hint of being outdated. Imagine such a production period for any car model!

Surprisingly, Rolex is struggling to maintain itself in the chronometer competition which has been ongoing since 1945. The leader here is its rival Omega. However, in 1949, the watch regulated by J. Matile won in the category of chronometer wristwatch with 849 points.

Wilsdorf suffered a stroke of fate in 1944, when his wife, May Florance, died. Since the marriage was childless, he transferred his shares in Rolex Montre SA to the Hans Wilsdorf Foundation; a wise decision regarding the continued existence of the company, as other companies, which function under the foundation model, have proven. Examples are the piston manufacturer Mahle or Bosch, both based in Stuttgart. From all these foundation-model enterprises, as also from Rolex, considerable financial resources flow out to scientific, social, or charitable projects around the world.

The 70th birthday of the popular patron was celebrated in 1951 with a four-day festival in Geneva. In fact, this festival was also to celebrate his 50 years in the service of horology, as well as the 25-year anniversary of the Oyster case and 20 years of the automatic winding mechanism. Despite his advanced age, Wilsdorf still determined the fate of the company, even if he now had two directors on hand. Such innovations such as the Cyclops date magnifier on the dial are the result of his innovative policy.

Until 1965, Rolex Montre SA operated out of cramped quarters on three floors on the Rue du Marché, while more of the 400 employees were spread out in neighboring buildings. At this time, some 250 people were working in Biel. Since the Great Depression of the 1930s and the consequent devaluation of the pound had made the English market a difficult terrain, Montre Rolex SA decided to internationalize. This was undertaken by subsidiaries worldwide, something Hans Wilsdorf personally directed himself in the 1950s. When he died in 1960 at the age of 79, it was the end of an era. Starting in 1963, his confidant André J. Heininger directed the company's fortunes as president of Montre Rolex SA and of the Hans Wilsdorf Foundation, until 1992, when his son Patrik took over management of the company.

Wilsdorf did not witness the company's move in 1965 to its new building on Rue Francois-Dussaud, but did, however, see the launch of the new specialized Oyster models, which began in 1953 with the Turn-o-Graph and the Submariner. These durable watches for daily wear, as well as for climbers, pilots, divers, oceanographers, racing drivers, scientists, Hollywood stars, and adventurers, still form the moral backbone of the brand. On another side, it caters, with precious metal sports watches set with diamonds, to some rather idiosyncratic consumer tastes. Sports legends such as Jean-Claude Killy, Jacky Steward, Arnold Palmer, as well as climbers like Sir Edward Hillary and his Sherpa Tenzing Norgay, and Reinhold Messner, and scientists such as Jacques Piccard have significantly influenced the image of Rolex as popular wearers of its watches. When the latter dove to 10,916 meters in the Mariana Trench in the Pacific Ocean, he was not wearing a Rolex Oyster on his arm, but rather on the outer surface of his bathyscaphe *Trieste*. After a successful diving record, Piccard confirmed the smooth functioning of the watch and thus gave all normal Rolex wearers the feeling that they had made the right decision when buying a waterproof watch.

Despite some models with full moon phase or full calendar, or even a Rolex chronograph rattrapante, Rolex remained for over a century true to its minimalist principles: No tourbillon, no chimes, no perpetual calendar, nor even a Grand Complication enhance the product's history. This has not damaged the brand's luster. Even so, Rolex has offered a variety of diverse and fascinating models throughout a century of watchmaking history, which have all been significantly influenced by the brand with the crown.

A James Bond Rolex. In Her Majesty's service, the agent timed the travel time of a cable car with this chronograph, to make sure how much time he had left to escape from his prison, suspended on the uphill steel cable.

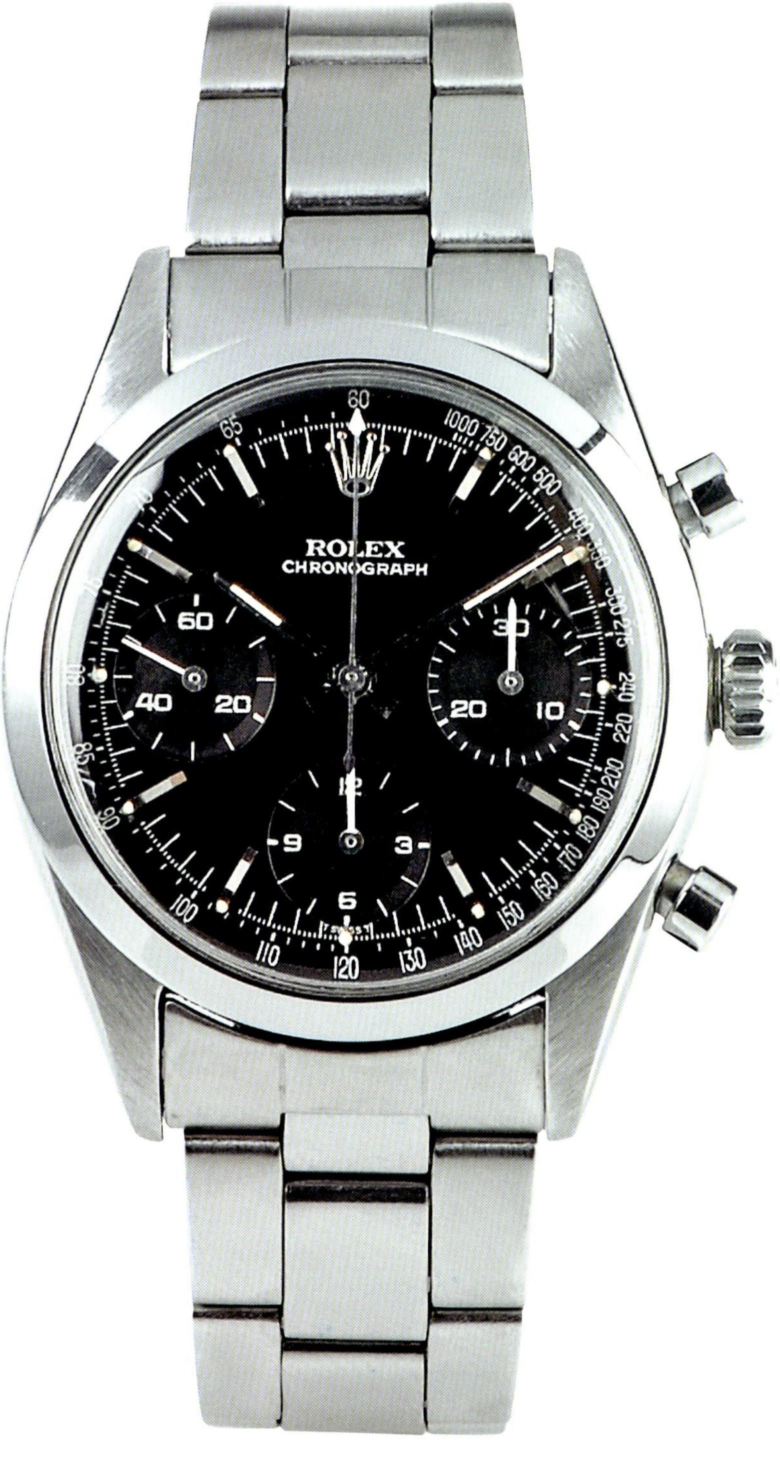
ROLEX
CHRONOGRAPH

ROLEX
OYSTER PERPETUAL
SUPERLATIVE CHRONOMETER
OFFICIALLY CERTIFIED
COSMOGRAPH
DAYTONA
UNITS PER HOUR

The Rolex Oyster—The Long Road to an Elegant Sports Watch

It is likely that no watch design has more influenced the image of the classic sports watch than the Rolex Oyster models. In first place comes the Submariner, then the GMT and Explorer whose different versions have shaped that new type, the Tool Watch: high-quality watches for the widest range of requirement profiles, that function reliably at great heights, at either low or extremely high temperatures, in deep water, or in extremely strong magnetic fields. With this came the chronographs that made it possible, in a world increasingly determined by speed, to capture even the smallest unit of time exactly.

The starting point was the chronometer, an extremely accurate and waterproof watch. Rolex understood early on how to win testimonials—be they from athletes, scientists, artists, adventurers or mountaineers—for the brand. The most successful auto racer of the 1930s, with a popularity comparable to Michael Schumacher's today, was the Mercedes-Benz company racing driver Rudolf Caracciola, who, of course, would appear wearing a Rolex chronograph at the starting line. Little wonder that Hans Wilsdorf was also a committed Mercedes-Benz customer.

Record-setting driver Sir Malcolm Campbell, just like golfers Arnold Palmer, Jack Nicklaus and Gary Player, wore a Rolex. Skier Jean Claude Killy and actor and amateur racer Paul Newman were synonymous for certain models among collectors. Rolex was there during the first ascent of Mount Everest in 1953, worn by Sir Edmund Hillary and Tenzing Norgay, and also there in 1960—as a custom-made version—on the outer surface of the bathyscaphe *Trieste*, penetrating to a depth of 10,916 meters in the Pacific with Jacques Piccard.

In addition to sports watches, from the start until today Rolex has also been offering an elegant, albeit minimalist, line of watches. The Rolex Prince with its dual dial, which features the second hand on a separate second dial below the hour and minute display, has been manufactured since 1928 in different cases. Like the Railway, in a side-stepped case, or the Brancard in a waisted case, this model was also made in

Reference 116518 was the first Rolex chronograph with in-house movement, which was replaced by the externally purchased but thoroughly reworked Zenith 400.

a variety of precious materials. The Caliber T.S. Ref. 300 was a shaped movement with dimensions of 16.90 x 32.70 mm. The 18-jewel movement with swan's neck fine adjustment was chronometer-regulated and had a protective bridge for the balance wheel. This distinguishes it from other Gruen Techno movements, also delivered to Alpina, which offered a similar model. This circumstance makes it easy for counterfeiters today to convert such models into a Prince.

However, Rolex also demonstrates continuity with this model. In 2005, the Geneva-based company introduced the contemporary interpretation of a chronometer-certified, manual-winding watch with a shaped movement. While early shaped-movement watches and chronograph and non-Oyster models were manually wound watches, Rolex very soon made the waterproof Oyster models with its Perpetual caliber. This bidirectional self-winding rotor was patented in 1933. Rolex automatics with their integrated bracelet lugs also differed in design from the previous soldered on wire loops. These automatic models owe their names to the bulge necessary to accommodate the rotor on the back of the watch: the Bubble Back. The Hooded Bubble Back, which came on the market in summer 1938, with covered lugs, was the first bicolor steel-gold model offered. The same is true for the Rolex-typical bracelet, which forms an optical unit with the watch, and up to the present day has shaped the effect created by the sports watch line. With the Day-Date, launched in 1956, Rolex designed a watch that could deliver the weekday and date—important information for a businessman—directly. Of course, there was also a ladies' Datejust, and Rolex invested its whole ambition in having this line of women's watches also certified as a chronometer—no easy feat for these delicate movements. After the Second World War, the successful sports watch line was launched in 1953 with the Submariner; this, however, was preceded by the Turn-o-Graph, with rotating bezel, advertised as the Rolex minimalist chronograph. Already by 1936, the firm had delivered manual-winding movements to the Florence-based company Panerai, for a diver's watch for the Italian Navy: the Radiomir. The first prototypes were even manufactured at Rolex. Thanks to the Panerai family's good relations with Hans Wilsdorf, the 17-jewel Rolex manual-winding movements were delivered until 1952. For the very diverse family of sports watches,

The new Rolex Prince is an extraordinary masterpiece among wristwatches. A transparent display case-back for the first time allows a view of a guilloched Rolex watch movement; the pattern also decorates the dial.

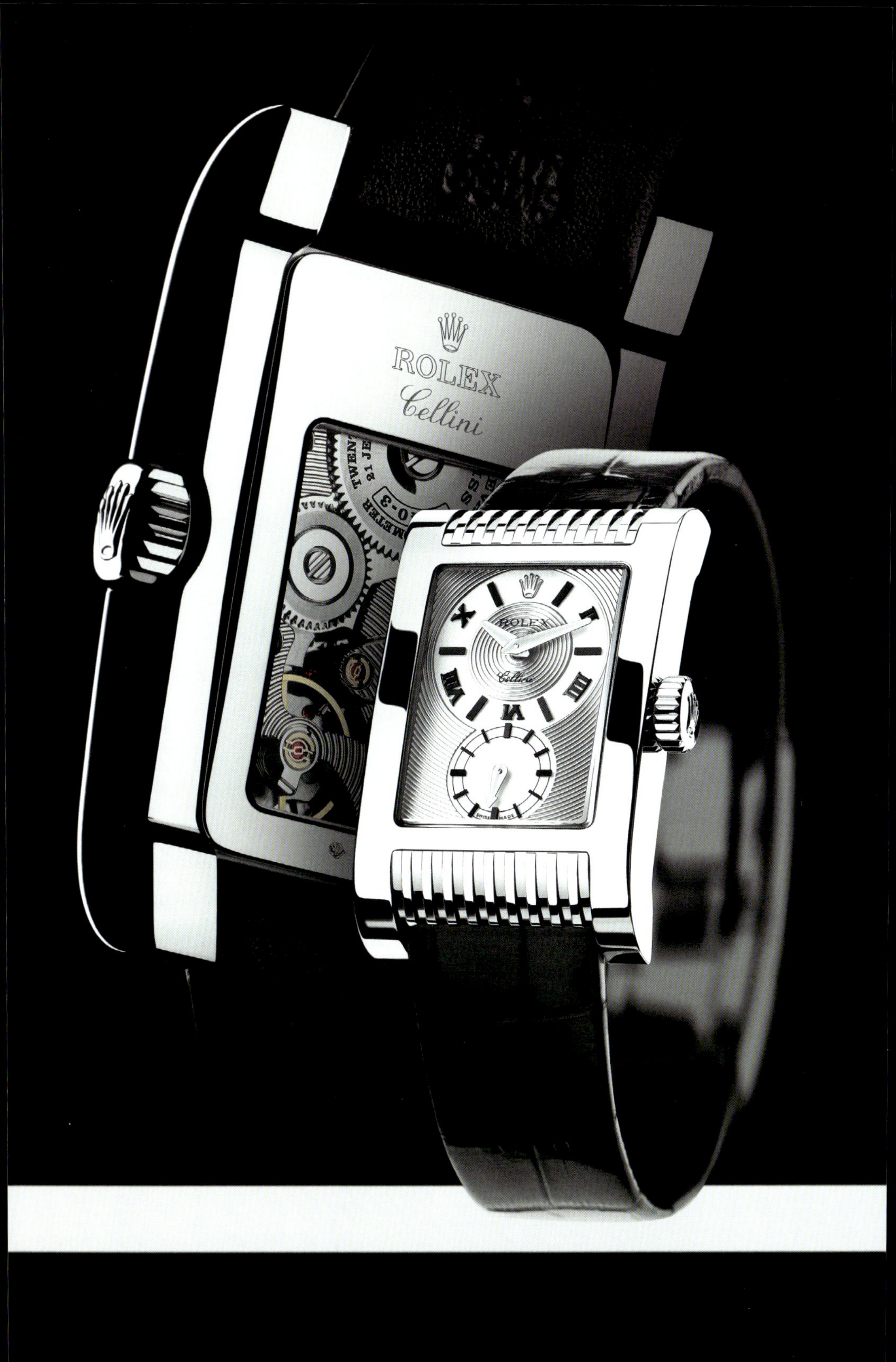
ROLEX
Cellini
ROLEX
Cellini

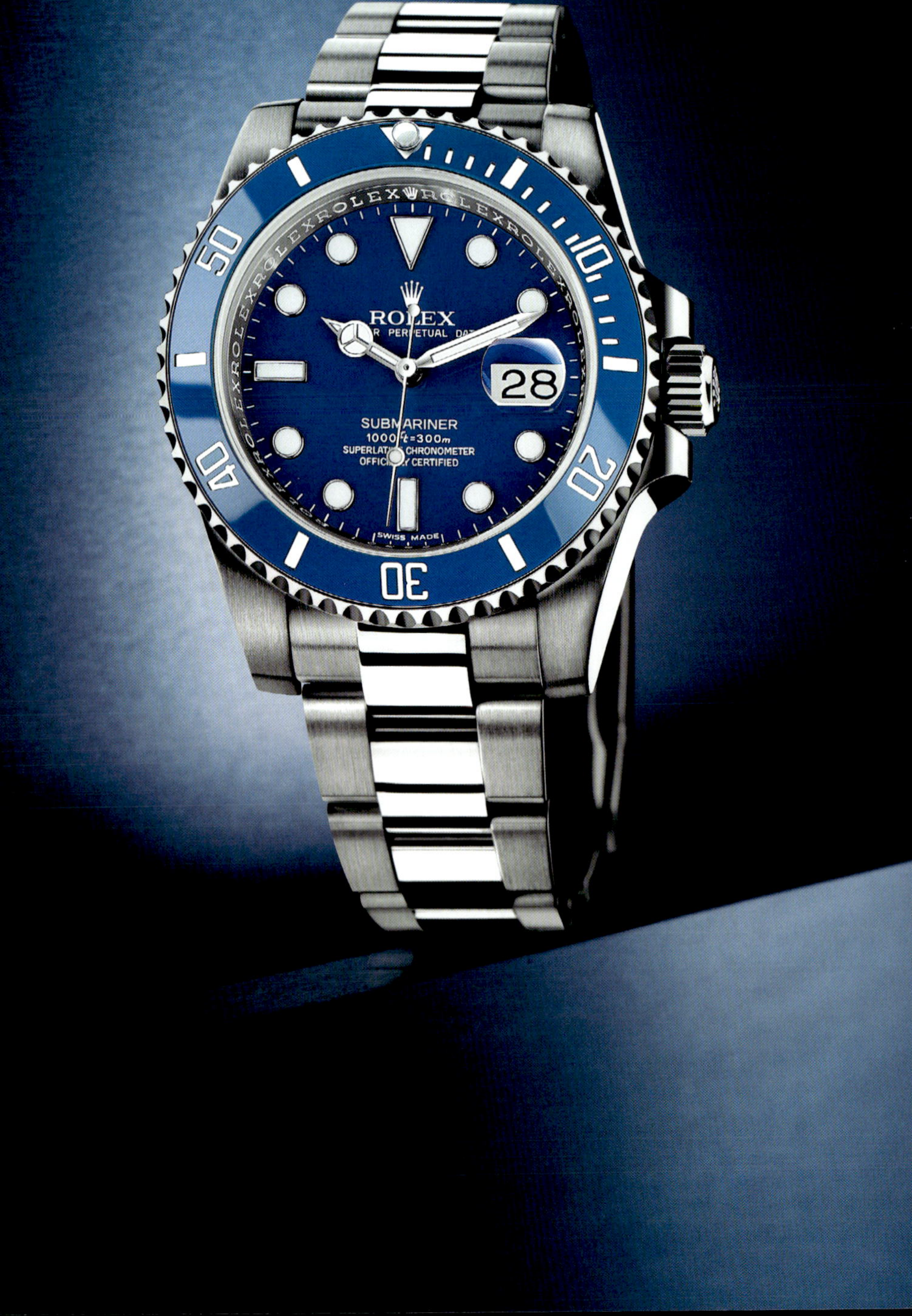
ROLEX
SUBMARINER
28
SWISS MADE

excepting the chronograph, a basic movement was generally used, modified according to the intended use, such as of magnetic field protection, a second time zone, 24-hour display, etc. This development started in 1950, with the Caliber 1030, which has a diameter of 28.50 mm and a height of 5.85 mm. The oscillation frequency is 18,000 vph. The movement has 25 jewels and a screw balance with regulator discs.

In 1957, Caliber 1530 followed, which, although it has the same diameter, is slightly flatter with a height of 5.75 mm. The oscillation frequency is the same (26 jewels, screw balance). In Calibers 1520 and 1580, derived from this movement in 1963, the oscillation frequency was increased to 19,800 vph.

With the Caliber 3035, which has a diameter of 28.50 mm and a height of 6.35 mm, a completely new movement was introduced in 1977. It has 27 jewels and a screw balance, with oscillation frequency at now 28,800 vph. The last stage in this chain is the Caliber 3135, presented in 1990. Rolex is just as conservative about its movements, as it is about the models in which the movements are installed.

In 1953, the Explorer is introduced in addition to the Submariner, followed by the 1954 GMT-Master and the Milgauss, which can resist magnetic fields up to 1,000 gauss. In 1961, Rolex presented its Cosmograph Daytona; in 1971, the Sea-Dweller 2000 (610 m); in 1980, the Sea-Dweller 4000 (1,220 m); and in 1983, the GMT-Master II. The Explorer II, especially designed for cave researchers, with a 24-hour display, also appeared in the 1970s. Some model designations, such as Precision, already existed for non-Oyster models. In 1992, Yacht-Master created a new product name, while the Air-King, an Oyster without date, was among the longest-manufactured models. This watch is also the most inexpensive alternative to wearing a Rolex Oyster.

Many of these models remain in the production program to this day—often indistinguishable from their predecessor models by a layman. Even the Milgauss, whose production ended in 1988, has been available again since 2007 in a revised version. Typical of most of these models is the Triplock crown with shoulder protectors. Because these models, designed as pure sports watches, are also available in various gold alloys and platinum, and since the Geneva watchmaker is not afraid to use precious stones to ornament the dial, bezel,

A diving watch in white gold. The Rolex Submariner is one of the watch legends that made the Tool Watch famous. Manufactured since 1953, it has remained almost externally unchanged up to today.

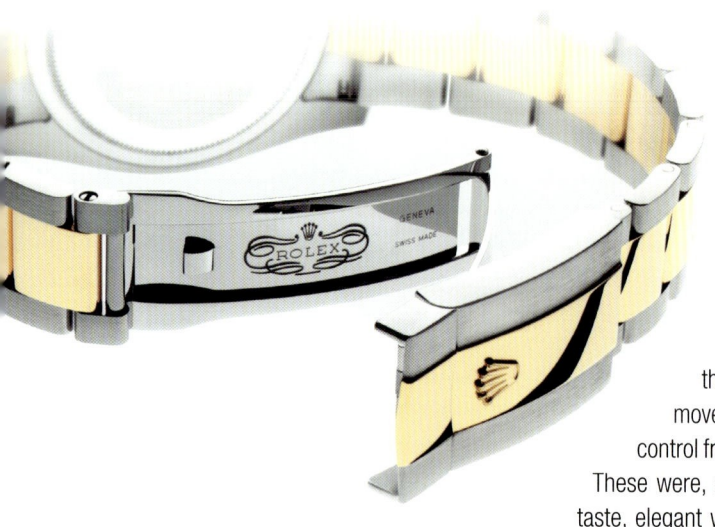

Often causing complaint because of their sharp edges, the simple clasp of the Rolex metal bracelet models is quite functional.

and bracelets, the house has gained a very special clientele, who, as the Swabians say, have *"ein G'schmäckle"*—a slightly shady character. But that does not detract from the quality of these watches, and in recent years the house has also made its gold Oyster models available as the serious customer likes to wear them: with leather strap.

Chronographs play a special role among Rolex sports watches. These are sought after not only as new watches, with the different models also achieve top prices at auctions. Anyone who bought a Daytona for $875 in 1978, can now—if it is in good condition and provided with appropriate purchase documents—sell it for twenty times the former price. The early Rolex chronographs were delivered as single push-piece chronographs, with the 13-ligne manually winding movement VZ with column wheel control from Valjoux starting in 1930. These were, according to contemporary taste, elegant watches with a diameter of 28-32 mm. Only a few exceptions, such as the Reference 2705 with a 14-ligne manual-winding Valjoux Caliber 22 movement, corresponded to the 37 mm modern day size requirements. With the replacement of the Valjoux Caliber VZ by the 22, both with small second at the nine and stopwatch minute hand at the three, the case size also increased. Besides the traditional round chronograph case, square cases were also used, as well as the delicate Valjoux 69 with 10½ lignes. At 44 mm, an impressive size even by today's standards, the Rattrapante Chronograph with the 17½-ligne Valjoux 55 VBR created the Reference 4113. After Valjoux 22 and 23, followed the Valjoux 72 in 1948, which was to power Rolex chronographs and all their derivatives, from the calendar watch, the Cosmograph, up to the latest manual-winding Daytona of References 6265 and 6263. In the 14 or 18 carat gold version, the watch was offered, with the Caliber 727, as a certified chronometer for the first time. The movement was used as the Valjoux 72 C in the so-called pre-Daytonas. The calendar function displayed month, weekday, and date. The Valjoux 88 added a moon phase. Reference 81806, commercially available at the beginning of the 1950s, was made only in very small numbers, and is now exorbitantly expensive, although the movement was made by other manufacturers. With a case diameter of 40 mm and a 13-ligne automatic movement based on the Zenith 400, Rolex introduced the first automatic chronograph in its history in 1988. In contrast to the Zenith El Primero, the Rolex Caliber 4030 oscillated at 4 Hz

Chronograph Caliber 4130. A self-winding column wheel caliber , which powers all Rolex chronographs today, and in 2000 became the first independent chronograph movement presented by Rolex.

(28,800 vph) instead of at 5 Hz (36,000 vph). All models were certified chronometers; the bezel was polished and engraved. In the first models, the tachymeter scale ranged from 50 to 200 km/h, later on, from 60 to 400 km/h.

Although the modified Zenith Caliber was a functional and visually successful movement, and due to all the various modifications could be considered an in-house product, it was Rolex's ambition to also become independent of other manufacturers in the chronograph sector. Therefore, in 2000 they presented the Caliber 4130, a completely in-house design. The column wheel controlled chronograph was made flatter than its predecessors and also has a larger power reserve, which now is 72 hours. As in the Piguet 1180 movement and its derivatives, Rolex decided on a vertical friction coupling for its new design; this prevents the second hand from jumping when started and allows for continuous wear-free operation of the chronograph. The only visual difference is that the permanent small seconds is at the six.

For collectors of Rolex Tool Watches, things are the same as among philatelists: marginalia can determine differences of several thousand dollars or euros. A manual-winding Daytona with a seemingly empty dial, known among collectors as a "Paul Newman" because he once won this watch as a prize in a race and was then photographed on the cover of an Italian magazine, costs at least twice as much as its technically absolutely identical sister models. Among the pre-Daytonas, it is the Reference 6236, called the "Jean-Claude Killy" by collectors, with full calendar and steel bracelet, that makes hearts beat faster and achieved up to $110,000 at auction. These models were waterproof to 50 meters. This was only noted on some models as "50 m = 165 ft."

For Submariners, earlier models without crown protection are increasingly in demand. This watch became popular after being worn on the wrist of Sean Connery in the film *Dr. No*. The Submariner logo in red also earns a significant supplement. If it is a Sea-Dweller with a Comex imprint and the watch has the appropriate documentation, the price can break the $100,000 barrier. It is clear that models such as the Milgauss, made up till 1988 and now re-issued, wins top prices. Likewise, the imprint "Tiffany & Co." or "Cartier" on the dial consistently increases the value. A Submariner with a blue-gray bezel instead of black, a "Paul Newman" with red dial, known to insiders as the "Spirit of Japan," an early Turn-o-Graph, are all—provided they are authentic—models that achieve extremely high prices. Technical changes can also contribute, as in the manual-winding Daytona. With a Caliber 727—the basic caliber here is also a Valjoux 72—the oscillation frequency increased from 2.5 (18,000 vph) to 3 Hz (21,600 vph). This consistently increases the value of the Reference with Caliber 727, but it was indeed the last in a long line of Rolex manual-winding chronographs.

Collectors take note of even the little things—such as if the depth indicator on a Submariner is in feet or meters first—and

Rolex offers today a line of elegant watches in the Cellini model range, along with its sports watches.

they differentiate between the "feet first" and "meters first." One watch was denied celebrity status, even though it played a supporting role in a James Bond movie. It is the Reference 6238, a chronograph with steel bracelet and silver dial, which Bond wore in the movie *On Her Majesty's Secret Service*. The reason this beautiful chronograph did not become a cult Bond watch may be the fact that it was not Sean Connery who played the Bond character, but the relatively unknown Australian actor George Lazenby.

Undoubtedly, the trend of rising prices at auctions for these sports models will continue. Special versions of the Daytona, as well as the Submariner and Milgauss models, regardless of their materials, exceed the $100,000 limit, and there is no end in sight to this tendency.

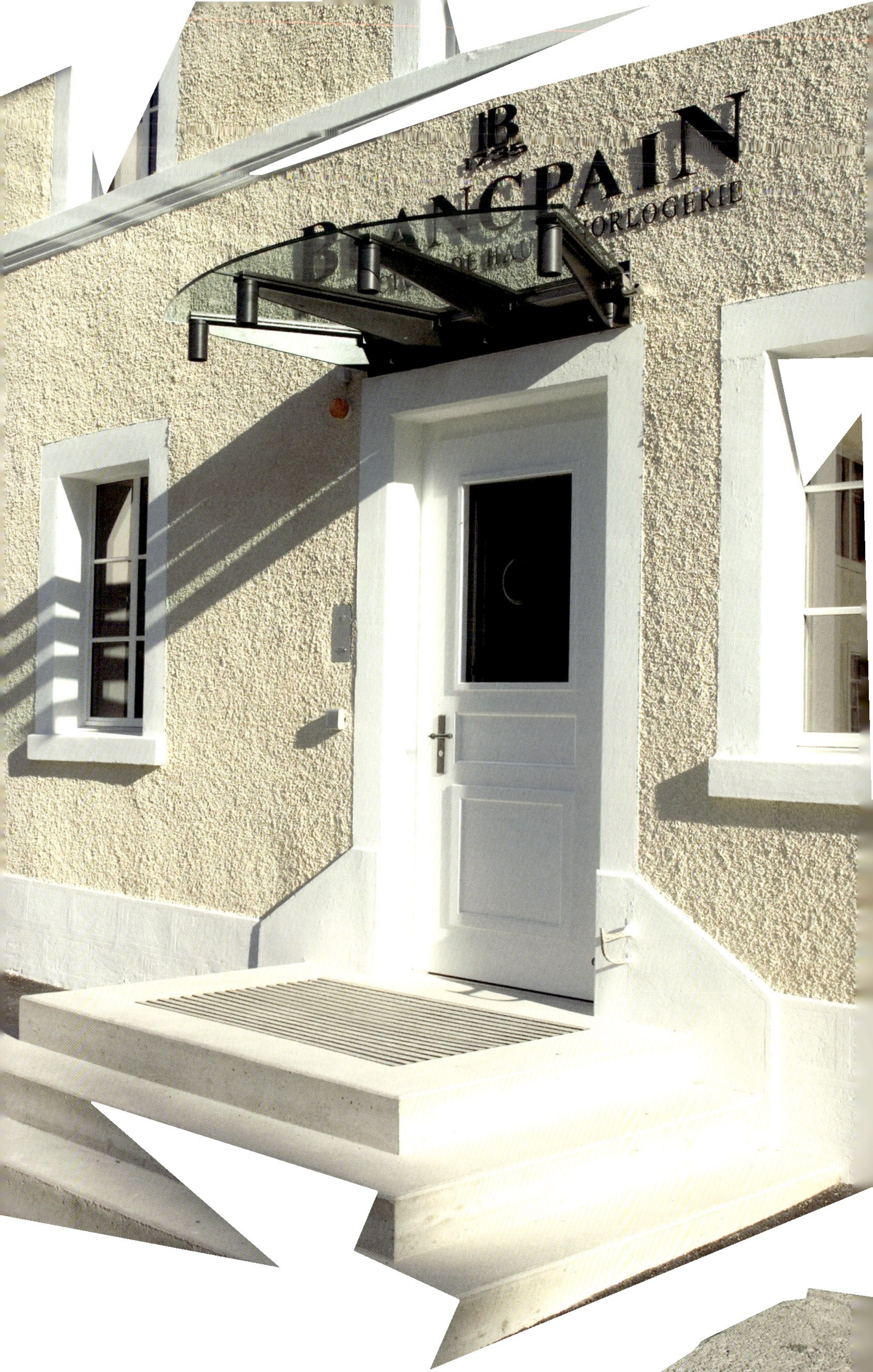
B
1735
BLANCPAIN
HORLOGERIE

Watches since 1735—Blancpain, a Great Marketing Story

With the slogan, "Since 1735, there has never been a quartz Blancpain watch. And there never will be!"—the revitalzied Blancpain watch brand re-entered the international arena of *haute horlogerie* in 1983. With its radical commitment to the mechanical watch and legacy of savoir-faire, they deliberately made recourse to the culture and tradition of craftsmanship that characterize the traditional craft of watchmaking. The credo: watches are works of art that receive their soul from the workmanship of people working consciously in a traditional context. Whoever seizes on such a watch, obtains more than a timepiece anyway available everywhere. It is much more a piece of craftsmanship and culture.

These are the considerations expressed especially by the convictions of a young watch industry manager, who believed in the long-term survival of the mechanical watch. Jean Claude Biver left his job at SMH with the Blancpain trademark in his luggage, which he had bought for CHF 20,000. Together with his friend and business partner Jacques Piguet, son of the blank movement maker Frédéric Piguet, he brought out a series of purist watches with his own movements. The production site of the new Blancpain S.A. Fabrique d'Horlogerie was Le Brassus in the Vallée de Joux. Thus, life returned to a watch brand that had quietly disappeared from the market back in 1970. It was then part of the SSIH, which also included Omega and Tissot. Since 1992, Blancpain has again been part of the Swatch Group, where many leading watch brands have found their new home.

In 1735, Jehan-Jacques Blancpain assembled his first watch as, so to speak, a sideline watchmaker. At the end of the 18th century, one of his sons, David-Louis, began to export the watches abroad. Another of his sons, Louis-Frédéric, then founded the watch factory in the city of Villeret, which (with a series of changing names, in the end called E. Blancpain & fils) was owned by the family until 1932. They made the first prototypes of a functioning automatic watch for John Harwood in 1926, and then as a concession, a series for the French market.

They also made the Rolls model for the Parisian watchmaker Hatot. When Frederic-

Still operating in a farmhouse that it has occupied since 1735, as the legend goes. No brand understands better how to convey savoir-faire.

Emile Jr. died in 1932, he left no descendants. The Blancpain line ended. His longtime confidante Madame Fichter took over the business for the next 40 years. Under the name Rayville S.A. succ. de Blancpain, such interesting watch models as the Fifty Fathoms were manufactured. Jacques-Yves Cousteau wore this diver's watch on his expeditions and also in his film *The Silent World*. On the other hand, the company made a petite ladies watch of 5 lignes (11.85 mm), thus manufacturing the smallest womens' wristwatch made at the time.

During the 1980s, Blancpain initially focused on petite, round, precious metal or steel watches with traditional complications, such as moon phases, perpetual calendars, or a chronograph with the smallest automatic column wheel caliber of its time. Then, in the 1990s, Biver took an about-face to make masculine, durable sports watches, yet fully suitable for everyday wear. The 2100 model range guaranteed being waterproof to 200 meters and even offered a waterproof tourbillon and chimes in a new 38 mm case design.

To maintain the level of the other luxury brands, Blancpain also produced a Grand Complication. This platinum watch, inconspicuous at first glance, for a long time represented the reference model for what was micromechanically feasible. With a split seconds chronograph, perpetual calendar, tourbillon, minute repeater, and automatic winding, the watch offered essentially everything that connoisseurs wanted. With a price that then approached $700,000, this masterpiece is also in the highest range. The basis for this watch, made of 350 individual parts, with 13½ lignes, is the Piguet Caliber 33, as are other watches with specifically developed Piguet movements. Only 30 pieces are planned. But the message is clear: we have been making watches since 1735, and we play in the top league.

In the small volume *The Ethics of Blancpain*, we can read: "The Blancpain watch is a piece of puritan art taken to the extreme. These are watches which express only the essential: the art of watchmaking. A watch which is freed from everything superfluous, and left the most important thing, the essential. As in the drawings of Matisse. The case is designed to follow the contours and dimensions of the movement as closely as possible. The idea is to bring out the genius of the watch to its fullest advantage. Because, if you take a very complicated movement and envelop it in an exceptionally restrained form, nothing will keep its oscillations back from you." As a watch enthusiast, one could only agree with this poetic philosophical reflection. Nothing is more inorganic than a tiny movement in a watch with a 45 mm case diameter. A good example is Blancpain's beautiful chronograph.

Luxury watches were traditionally made either of gold or platinum. Steel remained for use in more commercial products. In addition to Patek Philippe, Blancpain has made steel a socially acceptable material for luxury watches, and, for some of these pieces, you must pay many times as much as for platinum or gold watches from less

Watch movements can be productions to captivate the viewer. This is also the intention of Blancpain's "only watch."

reputable manufacturers. The same is true for the Blancpain pilot's chronograph, a watch in which the internal values count: for example, the winding rotor of 18k gold. In a chronograph, the movement construction plays a significant role. The connoisseur wants a column wheel control. The Piguet 1185 in this watch is one of the movements sold only in small quantities to reputable manufacturers. Because of its modular design, it can be manufactured as a manual-winding, rattrapante, or flyback, and is often mistakenly advertised as a modular caliber. This is clearly incorrect. It's a thoroughbred, highly exclusive automatic chronograph movement. Structurally, it is also the smallest and flattest movements of this type. It is an addition timer with flyback function. This design principle was primarily used for military aviation chronographs. In the German *Bundeswehr*, or army, the Valjoux 230 powered the Heuer or Leonidas chronographs. The special feature of this quick reset is that the stopwatch can be activated immediately without interruption by pressing the lower push-piece. This eliminates the time-consuming stopping and resetting the chronograph hand.

The limited edition of the Le Brassus Quantième Perpétual GMT combines three complications very useful for everyday: perpetual calendar, second time zone, and a moon phase at the six. The dial displays the date at the three, weekday at the nine, and the month at the twelve. All displays automatically take into account the different lengths of the months in the annual cycle, as well as the leap years. All these displays can be conveniently adjusted with correctors set under the lugs.

However, these watches are not only at the leading edge in terms of their movements. Both the leather strap design and the matching steel bracelet meet the highest standards. All solutions are lovingly detailed for Blancpain watches and well thought out. It starts with a simple, yet very elaborately crafted calfskin or alligator leather strap, with its underside stitched with Neoprene. The clasp is cleanly made and features the Blancpain logo. In this design, the watches have an unbelievably subtle look. Some models are more striking, with alternating polished and satin-finished steel or precious metal bracelet. The lugs are milled from solid aluminum. The bracelet is fastened with a solid butterfly clasp. The chronograph weighs 61 grams with leather strap, and 141 grams with steel bracelet. For water sports lovers, the latter version is recommended, because the screw-down crown and screw-down push-pieces guarantee water resistance to a depth of 100 meters. The execution of the screw-down crown and push-piece design is exceptionally well done. The crown threads grip by suction, and the push-piece screws expand the pusher contact surface when the screws are tightened

The flyback is a feature of the 2100 model range. This was launched by Blancpain to create a watch both luxurious and fully suitable for everyday wear. All complications that Blancpain has in its program are now available in the new case designs.

The domed sapphire crystal has double anti-reflexive coating, so that, especially in daylight, you get the impression that there is no crystal in the watch. The dial is black and the hands, like the Arabic numerals, are luminous, so the watch has excellent readability at night. The hands have become very sweeping, and this brings us to the only criticism of this watch: at times they cover the totalizers, so you have to wait a moment to read the marked time. The setting of the totalizers and small seconds is classic on the Caliber 1185: the minute at the three, the hour at the nine, and the small seconds with integrated date display at the six. Of course, the date display has a quick reset activated by pulling the crown twice, so even after extended wear, the watch is quickly re-adjusted.

Today, Marc A. Hayek, grandson of Nicolas G. Hayek, directs the company that, since 2005, has had its entire production in Le Brassus. He continues its successful concept under the slogan "Blancpain—tradition of innovation since 1735." Classic watches, based on such traditional series as the Fifty Fathoms are being newly interpreted. Proof of the company's competence, was newly reaffirmed by the Apotheosis Temporis collection, which is available only as a set of eight watches. These watches are now all self-winding and delivered with a watch winder, which ensures it is always ready for its fortunate owner. The set includes a tourbillon, an extra flat watch, minute repeater, a single push-piece chronograph with split seconds, a moon phase watch, perpetual calendar, time zone watch, and a perpetual calendar with sun hand and elliptical disc which "programs" the solar time. All these watches are set in a 38 mm case made of platinum. Another highlight of the new pieces is the perpetual calendar, with the corrector push-piece for the displays hidden under the lugs. These can now be operated without tools—a real step forward in a complicated watch.

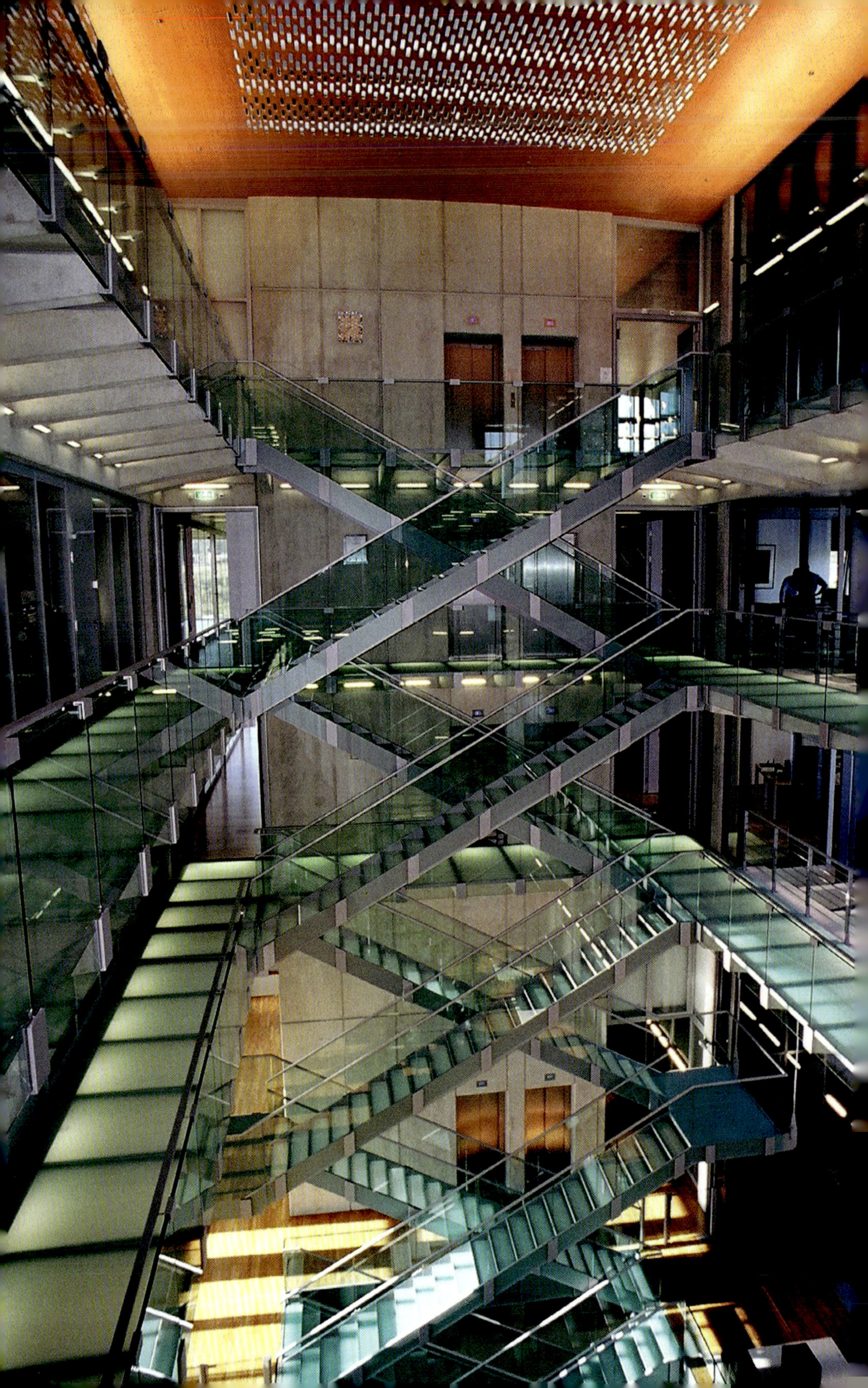

From 1775 to Today—Vacheron Constantin, Oldest Watch Brand in the World

Vacheron Constantin, with its birth year of 1755, is demonstrably the oldest continuously manufacturing watch brand in the world. The 24-year-old master watchmaker Jean-Marc Vacheron started the brand in Geneva. From generation to generation, the company was handed down in the family, until great-grandson Jacques-Barthelemy Vacheron followed up their previous collaboration by offering his friend Jean-Francois Constantin a partnership. Thus, on April 1, 1840, the Vacheron Constantin trademark entered the Commercial Registry. When Georges-Auguste Leschot was appointed technical director of the company in 1839, manufacturing processes were reorganized and serial production began. This was made possible by the pantograph, a machine that allows maunfacturing exact duplicates, creating parts that were interchangable, something previously not possible for purely hand-made products. At the same time, the company changed its headquarters and moved to the Tour d'Ile at the edge of Lake Geneva in the city center.

The company is a recognized achiever in Geneva and received an award for the best pioneering work in the field of watchmaking from the Société des Arts de Genève.

When César Vacheron died in 1869, his son Charles took over the company. When he lost his life only a year later, at the tender age of just 25, the two widows, Catherine Etierinette Vacheron, then already advanced in years, and Laure Vacheron Pernessin, took over the company. It had had a number of name changes in the past few years, but as Jean-François Constantin returned to the company in the mid-1870s, the company again became Vacheron Constantin.

In 1880, he registered the Maltese cross as the logo of the Vacheron Constantin brand. The inspiration for this was replicating a small wheel on the watch barrel cover.

After the death of the two ladies, the company was converted into a corporation. In 1911, it began to manufacture wristwatches, while the development of complicated watches moved forward simultaneously. Vacheron Constantin

Goethe said: "People need roots and wings." For a traditional company, that also means developing new ideas. At Vacheron, these include translucent sapphire crystals with tamper-proof secret signature.

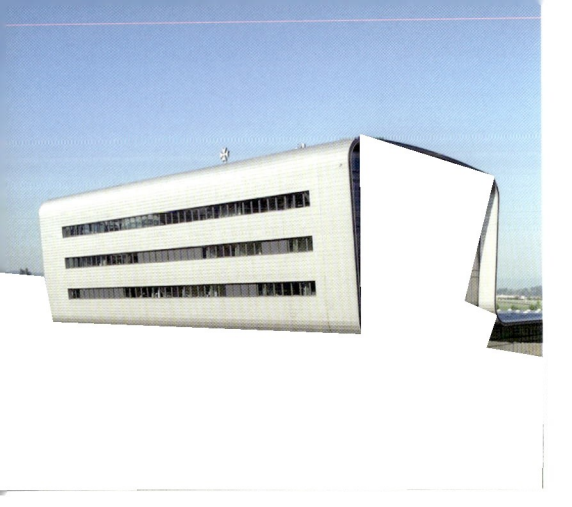

The new factory and administration building in Plan-les-Ouates, near Geneva, the new home of Vacheron Constantin.

mastered all the special features of *Haute Horlogerie*. It made a Grande Complication for King Farouk of Egypt, one of the most complicated watches of that period. In the same year, 1938, collaboration with Jaeger-LeCoultre was intensified, including use of their blank movements. But against the backdrop of World War II, business was bad for a luxury watch manufacturer. In 1940, Charles Constantine was forced to sell the majority of shares to Georges Ketterer, who became the main shareholder of the company.

The Ketterer family sold their majority share in 1985 to the former Saudi oil minister Sheikh Yamani, who, however, took no direct role in management of the company, which remained in Swiss hands. Today, Vacheron Constantin is part of the Richemont Group and employs approximately 400 people worldwide, most of whom work in the factory in Plan-les-Ouates, near Geneva. This Group also includes Cartier, Piaget, Panerai, and Mont Blanc, and the trademarkss acquired by the former Mannesmann Group, Jaeger-LeCoultre, A. Lange & Söhne, and IWC. Vacheron Constantin has representatives in 80 countries around the world and is sold in over 15 of its own boutiques and by a network of approximately 500 outlets. In 2005, the company celebrated its 250th anniversary. The Malte line (this name recalls the Maltese cross, the hallmark of the House) is the backbone of the Vacheron Constantin range, and distinctly highlights its strong points. In these timepieces, modern design and highest-quality watchmaking technology merge to create their own strong identity. The Malte Collection includes chronographs with power reserve and date, minute repeaters with perpetual calendar, a regulator with dual time display, a perpetual calendar with retrograde display, as well as a tonneau style chronograph, dual time, and several tourbillion versions.

The Patrimony Collection gladly shows its classic qualities and embodies the core

One of the Patrimony Collection models with manual winding and small seconds. Like most models, this one also displays the Hallmark of Geneva, which ensures the highest level of watchmaking quality.

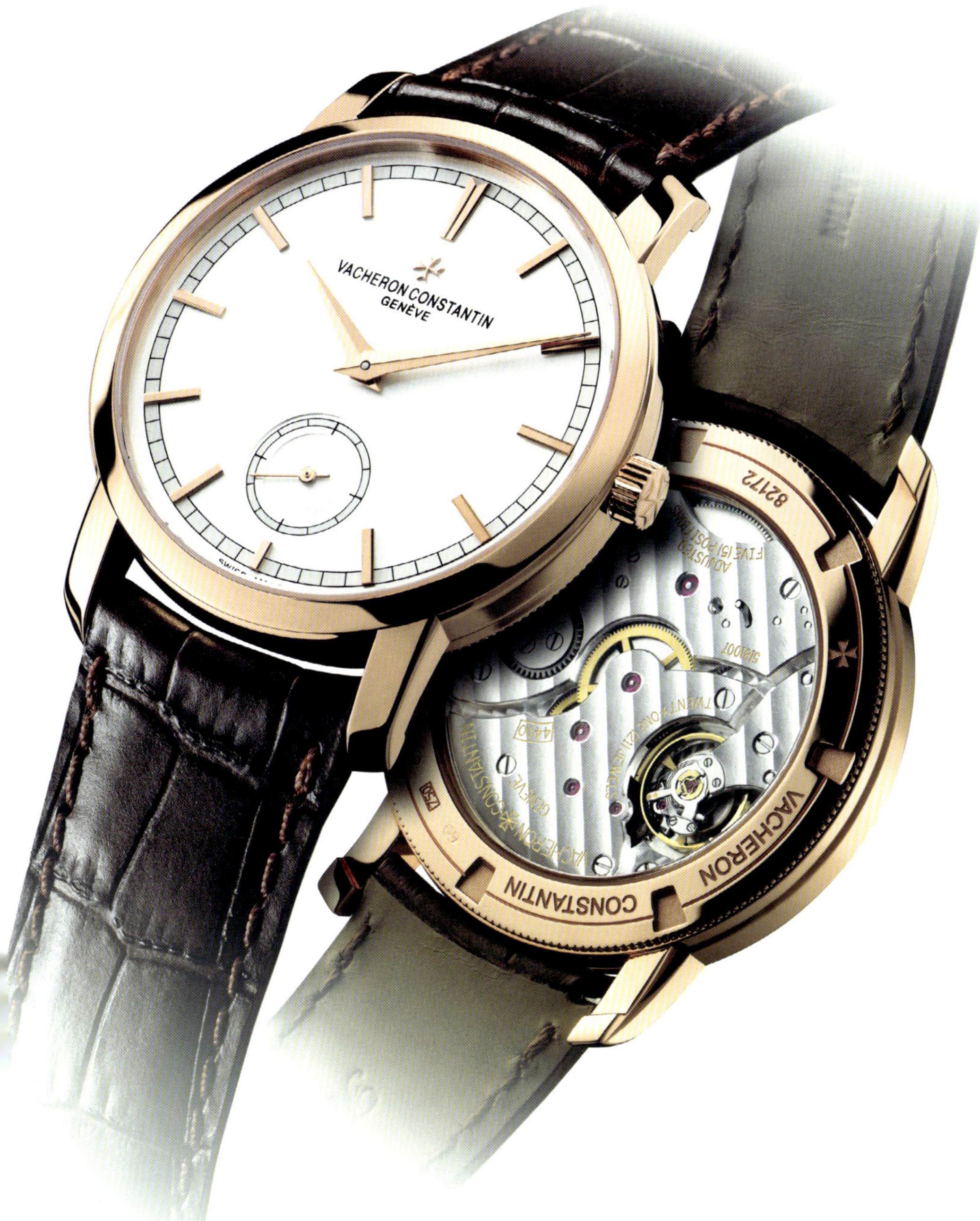
VACHERON CONSTANTIN
GENÈVE
82172
VACHERON
CONSTANTIN

values of Vacheron Constantin. The Patrimony models, keeping time with movements ranging from simple to the most highly complicated, offer elegance with contemporary style, combined with the unique know-how and experience of Vacheron Constantin. They are available in the broadest range of styles, as Classique with or without gold bracelet, as the Contemporaine model, an automatic watch with date, and as the Traditional model with perpetual calendar.

The Overseas is a sporty-technical line. It comes, in addition to the three-hand model, as a chronograph and dual time watch with second time zone display; beyond this is the Ligne Historique, which, inspired by iconic pieces from the House's rich heritage, gives rise to creation of new interpretations. Currently, there are two models, the Chronomètre Royal of 1907 and the Toledo of 1952.

The Métiers d'Art line of watches unites the expertise and attainments of centuries-old craftsmanship: the enameling, jewelry work, skeleton movements, hand engraving of the movement parts and cases. In this range of watches, you will find watches set with brilliants, as well as enameled dials

Côtes de Genève decoration and perlage, even where you don't see them, distinguish these from other high-quality watch movements.

with world map designs. The Cabinotiers Collection reaches the pinnacle of high-quality horology. The most creative talents at Vacheron Constantin work on these watches, to create true masterworks, such as the skeleton minute repeater, equation of time, sunrise, and sunset. Even after more than 250 years, this is an impressive brand with fascinating products that will carry on in the spirit of its founding fathers.

The Grande Complication, originally made exclusively for King Farouk. Note the power reserve indicator for the Sonnerie—the watch's chiming mechanism—at the nine. The power reserve display for the movement is at the three.

TACHYMETRE
HEUER
CARRERA
THU 8
CALIBRE 16
100 METERS
CHRONOGRAPH
AUTOMATIC

Fascinated by Speed—TAG Heuer, the Racing Driver's Watch

No watch company is so closely connected with the chronograph as Heuer. From the 1940s until well into the 1970s, it was the desire of every young man to call a Heuer chronograph his own. The Carrera with the first Automatic Caliber 11 or the Montreal with the same movement, exercised a tremendous fascination on sports fans and racing enthusiasts. Popular wearers, such as Steve McQueen, Niki Lauda, Clay Regazzoni, or earlier on Juan Manuel Fangio, made this watch the instrument of professionals. Even Enzo Ferrari, ever critically confrontationist in regard to advertising on his cars, allowed Heuer stickers on his Formula 1 racing cars. Thus, Jody Scheckter, Jacky Ickx, Mario Andretti, and Gilles Villeneuve became the men who wore the Heuer watch advertising message on their wrists, while timing for the Scuderia Ferrari was done and further developed by Heuer specialists. From 1971–1979, Heuer was Ferrari's official timekeeper. In addition, in the 1950s and 1960s, German *Bundeswehr* pilots wore Heuer flyback chronographs, which have a Valjoux Caliber 230. In its manual-winding watches Heuer primarily used high-quality Valjoux calibers. Frequent flyers proudly wear their Heuer Autavia GMT with a modified Valjoux 72, a manual-winding column wheel movement with 24-hour display.

However, like so many others, Heuer also fell into the "quartz trap." Initially they positioned themselves quite successfully in the market, with such high-priced models as the Chronosplit GMT with digital and analog display, or in 1975, with the first quartz Chronosplit chronograph with LCD Display—in addition to continuing to manufacture mechanical models. But then prices for quartz technology went through the floor, and more and more, cheap watches from the Far East dominated the market. While the Microsplit, launched in 1972, as the first watch able to measure the tenths and even, a year later, the hundredths of a second, was initially sold in a decorative red box for 1,500 CHF (US$400 in 1972), four years later, the price was just 100 CHF (US$33). At these prices, the company could hardly realize the high development costs.

With the Heuer Carrera Caliber 16, the company designed a glorious product, although the modern interpretation features a modified Valjoux 7750 automatic movement.

Edouard Heuer had launched the company in 1860 in St. Imier. Nine years later, he received a patent for a keyless winding mechanism. As early as 1882, four years before the invention of the automobile—which increased the pace all over the world—he started to make pocket chronographs. When Edouard died in 1892, Charles-Auguste Heuer and Jules-Edouard Heuer took over as sole shareholders at Edouard Heuer & Co. By 1902, the company was making a turnover of CHF 152,000. In 1912, it was renamed the Ed. Heuer & Co., Rose Watch Co. and commenced making ladies' watches.

First and foremost, however, it was concerned with the instrument which an ever-faster-paced society urgently required on land, on the water, and in the air: the chronograph. In 1914, the first wrist chronograph with a Valjoux movement and the winding crown at the twelve was manufactured. Two years later, the micrograph was presented, which can mechanically measure the 1/100th of a second. The company expanded, and bought the Jules Jurgensen brand in 1919. A chronograph of this make was used a year later as the official timekeeper at the Olympics in Antwerp.

The world economic crisis also left its mark on Heuer, and 1932 was the worst sales year in company history. The company bethought itself of its basic values, divested itself over the coming years of previously acquired companies. In 1933, it manufactured the first on-board clock for the then-popular long-distance trekking and off-road driving expeditions. In 1939, it proudly presented the first waterproof wrist chronograph.

The chronograph's potential became apparent after the war. In 1950, the first chronograph with tide indication came on the market. In 1960, Heuer bought the Leonidas watch company and traded as Heuer-Leonidas SA. In the same year, the Carrera line, named after the long-distance rally, was launched. Jack W. Heuer, who today is the Grand Old Man of the brand, joined Heuer in 1958, and took over the majority of shares in 1961.

Many chronograph versions appear: the Carrera in 1965, with digital date display at the nine, and in the same year, the already mentioned Autavia. The Microtimer, an electronic timer accurate to 1/1,000 of a second, followed a year later and demonstrated its competence in regard to electronic timekeeping. In 1969, the first automatic chronograph caliber with micro-rotor, and the Monaco, the first rectangular, waterproof chronograph, were introduced, and Jack W. Heuer went public with the company in 1970. In the United States, he witnessed the rise of chip technology in Silicon Valley. Bo Noyce, founder of Intel and a big Ferrari fan, showed Heuer one of the first wafer fabrication operations, which produce watch chips for LCD watches. At that time, $1 million would have been enough to buy such a factory, but Heuer did not have the capital, and his Swiss friends, with whom he discussed the opportunity, did not understand what was going on in the United States. So it was not the analog digital watch which brought down Swiss watches, but rather digital watches from

Fairchild and National Semiconductor, that took over the market for mass-produced watches from one day to the next with their $19.95 watches.

In 1982, Piaget took over Heuer Leonidas SA, and Jack Heuer retired from the company. Three years later, Piaget, whose expertise was more established in the area of jewelry watches, sold the company to the international Techniques d'Avant-Garde (TAG) group. Thus, Heuer became TAG Heuer.

In 1999, TAG Heuer was sold to LVMH group (Louis Vuitton Moet Hennessy). Here, they again recalled their tradition and core competence of manufacturing fascinating sport watches to measure short time intervals.

Today, in addition to the Monza Caliber 36 with its column-wheel chronograph caliber, TAG Heuer also makes the Microtimer digital chronograph with a quartz movement developed by Heuer, the first wristwatch able to measure the 1/1,000th of a second. The company keeps presenting surprising interpretations of the chronograph theme, such as the Monaco Sixty Nine, a reversible watch with a flat manually wound movement on one side and the Heuer quartz movement on the other side, a stopwatch accurate to 1/1,000th of a second. Traditionalists can select the Classic line, Heuer's new interpretations of historical models like the Carrera and Monaco, which feature mechanical movements ticking inside. Some are even designed as tributes to racing celebrities such as the Targa Florio, which displays on its back the dates when Manuel Fangio won his five World Championships.

Heuer has remained a chronograph specialist using both the Valjoux 7750 and the Zenith 400 with 36,000 vph. The company not only made automatic watches, but also chronographs with quartz technology.

TACHY
MON SUN SAT
TUE WED THU FRI
CHRONOMETER
Paul Picot
No 795
SWISS
MADE

Paul Picot—Even a Young Brand Can Make Traditional Watches

The name might suggest otherwise, but: Paul Picot is a young brand, born in 1976 amidst of the "quartz shock." The sense of commitment was to traditional watchmaking craftsmanship, and the desire, quite against the prevailing trend, was to continue offering high-quality mechanical watches. The initiator of this new company, based in Le Noirmont in the Swiss Jura, and its commitment to traditional horology, is the Italian Mario Boiocchi, who had been working as a watch importer.

Paul Picot was one of the first companies to assemble watches using the Lemania Caliber 5100, otherwise more often used in an attractive luxury chronograph. The Submarine, as the watch was named, offered both good design and internal workings that collectors found interesting. The same applies to the Le Rattrapante 310, a calendar chronograph with moon phase and split-seconds control, with a movement based on the split-seconds chronograph Caliber Venus 179. In terms of price, this model competes directly with the absolute leading brands, such as the Patek Philippe calendar chronographs that feature a perpetual calendar. The Technikum model is rather less expensive, incorporating three complications no other manufacturer offers in that combination. Reason enough to look at these exotics in greater detail. This model displays day and date, as well as power reserve. The watch also features a rattrapante chronograph. The Paul Picot Caliber 8888 is C.O.S.C. certified, identifying the watch as a genuine chronometer. The movement is based on the Valjoux 7750, which is used by many manufacturers, but thoroughly modified by Paul Picot. Some 70 more mechanical parts alone were added. The movement has a diameter of 28.9 mm (13¼ lignes) and a height of 8.7 mm. The automatic caliber is wound by a 21 carat gold rotor; oscillation frequency, like that of the basic caliber, is 28,800 vibrations per hour. Unfortunately, the rattrapant mechanism with column wheel and pliers is beneath the dial, and cannot be observed at work through the sapphire crystal back secured

This split-seconds chronograph, with a re-engineered Valjoux 7750 ticking inside, demonstrated Paul Picot's mechanical competence.

to the watch with six screws. Nevertheless, this hand-decorated movement, with its blued and gilded screws and colimaçon or snailing finished bottom plate, offers a wonderful view.

The opulently designed, three-piece case is reminiscent of a valuable pocket watch. Round, baroque shapes dominate, including the domed sapphire crystal. With a diameter of 40.5 mm and height of 15 mm, the Technikum is hardly any tiny delicate watch, but rather a presence on your wrist. It weighs 85 grams. The lugs on the watch are screwed on; on the buckle, fastened with the usual spring bars. The onion-shaped crown has flank protection and is very easy to operate. The chronograph function operates in the usual way for an addition timer, the push-piece for the split-seconds hand is at the eight. The pressure point can be readily found on the easy-to-use pushers. The only flaw of this otherwise effective chronograph was that the minute hand sometimes moved slightly along with the stopwatch when it was activated; the hand is not powered directly.

The watch is also available with metal bracelet, making it appropriate for sportswear. Since it is waterproof to 5 atm, there is no need to take off your watch when you go to the swimming pool. The watch's solid silver dial is artistically guilloché. The minute track is divided into intervals of five; the hours are only available as pyramid-shaped indexes. The tachymeter scale from 60 to 1,000 km is on the rehaut. The blue weekday display is in the minute totalizer at the twelve; the corrector pusher is inset at the ten. The hours totalizer usually provided by the Valjoux Caliber 7750 has made way for a power reserve indicator which can display about 47 hours. If you wear this watch regularly, the power reserve indicator will barely leave the fully wound position. The permanent small seconds is at the nine; at the three, designed as a totalizer, the date is displayed by hands. The date is set with the first catch position of the crown.

As with the weekday display, there is also a small spot of color here in the red 31. The hour and minute hands are designed as Breguet hands, which fits well with the slightly old fashioned look of this watch. Due to the many displays, readability is naturally not as good as on a normal three-hand watch. However, the Technikum is an ideal everyday watch for anyone looking for something special. Taking a look at the watch's price—nothing will change there any time soon. In terms of the quality of the complications, the Technikum's split-seconds mechanism, however, gets high marks. The relative obscurity of the brand ensures the wearer an unobtrusive quality. For those who want to carry understatement to the extreme, this watch is also made of platinum or white gold. The Technikum is, however, only one model among a series of interesting watches by this company. There is the C Type, a titanium chronograph with a whopping 48 mm diameter, or the Firshire, a watch with tonneau case, regulator dial and power reserve. The Technograph means a chronograph with totalizer and small seconds are decoratively arranged in retrograde and crescent form on right and left side of the dial. With the Atelier Tourbillon, which has a five-day power reserve and a display for the same, the company is clearly demonstrating their view of the brand's position: at the very top.

The Lemania manual winding Caliber 1883 provides the base for a moon phase display.

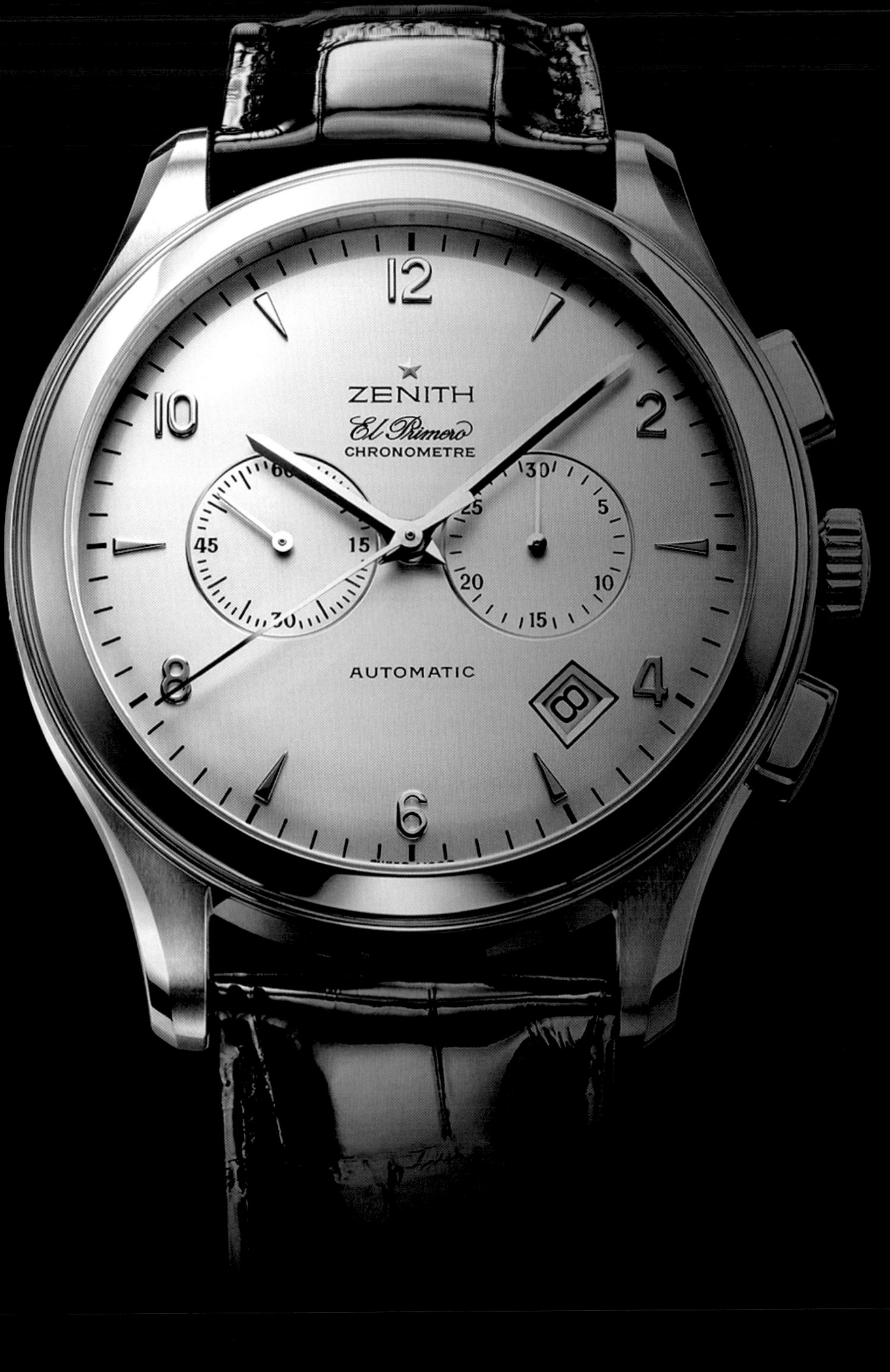
ZENITH
El Primero
CHRONOMETRE
AUTOMATIC

Now also at the Leading Edge in Marketing Technique—Zenith

Since 1865, Zenith has been making high-quality watches in the Swiss Jura town of Le Locle, first pocket watches and later wristwatches. It was the 22-year-old Georges Favre-Jacot who founded the *Fabrique des Billodes,* and at first, the watches were signed with his name. The company founder came to the product name of Zenith when watching the starry sky in the evening; it reminded him of the perfect meshing of the wheels of a watch movement. And so he called his watches after its highest point of the same: the Zenith. From these almost lyrical reflections on marketing, came the five-pointed star—up to the present, the hallmark of all Zenith watches.

At the turn of the 20th century, production was broadened to include board chronometers, table clocks, precision pendulum clocks, and later also marine chronometers. With the conversion of the company into a corporation in 1911, the brand now also became the company name.

The firm deserves credit for making the first automatic chronograph with central ball-bearing rotor. After five years of development work and shortly before the appearance of the Caliber 11 (credit for this goes to the project group of Breitling, Heuer-Leonidas, Hamilton/Buren and Dubois Depraz), Zenith proudly presented its own El Primero chronograph caliber with central rotor and oscillation frequency of 36,000 vph in March 1969. A classic design, which, as the Calibers 400 and 410 with full calendar and moon phase display, for years represented the ultimate in chronograph making.

Zenith stopwatches were the only ones that allowed timing to the 1/10th of a second. With a diameter of 13 lignes, which corresponds to 30 mm, the caliber is of an impressive size. The height is 6.50 mm, or 7.55 mm for the version with simple full calendar. A special feature is a classic column wheel to control the three chronograph functions of start, stop, and reset. This movement ensures its high level of precision, attested to by a chronometer certificate, by the glucydur balance, coupled with a first-class amagnetic and auto-

The El Primero Zenith Caliber 400, an automatic column-wheel chronograph, remains the basis for the most complicated Zenith models.

compensating anachron balance spring. The most complex version of the watch consists of 354 different components. It uses some 41 screws, and 31 jewels reduce wear. The simple version should be lubricated at more than 50 places. The oils used for this have ten different viscosities.

The extreme-stress high performance escapement receives four dry lubrications. The full calendar version comes with 38 lubrication points.

When the holding company, trading as Mondia-Zenith-Movado, was taken over in 1971 by the American Zenith Radio Corporation of Chicago (America's largest concern producing electronic components), the company soon came to serve only as a distribution base for U.S.-made quartz movements. By 1978, the Americans arranged that production of mechanical movements would cease and all existing movements, furniture, and machinery were destroyed. It is thanks to the insubordination of the then-head of the chronograph studio Charles Vermot, that this madness was not carried out. He hid large quantities of movements, tools, machinery, and all the design and production drawings for the El Primero Caliber in the firm's attic, with no one the wiser. This fascinating watch movement played a role in making the mechanical chronograph socially acceptable once again.

In 1983, Ebel presented a new chronograph using this movement, at a time when mechanical watches would just sit there on the store shelves. It has a quirky, almost playful design with roman numerals and a tachymeter scale at an angle to the dial. Despite its height of 11.2 millimeters, it appears very flat on the wrist because of the tapering back. A typical Ebel feature: the five screws securing the bezel. The Ebel dial lies beneath an anti-reflective sapphire crystal; the back fastens with a snap cover. Many other manufacturers also use the Zenith movement. Their most prominent client was Rolex, which, up to 2000, installed this movement in its chronographs after re-engineering reduced the oscillation frequency to 28,800 vph. Zenith also made beautiful chronographs with this movement, but maintaining a more conservative appearance, with the result that only insiders would recognize the true value of the brand and its technical potential.

In 1994, Zenith again presented an in-house caliber, the Elite movement. It took five years of research in Le Locle to develop this ultra-thin caliber. The movement is only 3.28 mm thick and reached 28,800 oscillations per hour. It has special technical features such as instant date reset, stop-seconds, more than 50 hours of power reserve, and fine adjustment. In the Zenith collection, this caliber is the one with the potential for the most combinations. In addition to hours and minute display and small seconds at the nine (one of the hallmarks of the brand), such additional functions as Dual Time and a power reserve indicator can be integrated. It comes as a manual-winding or automatic movement. The Elite movement made it possible to incorporate quality, reliability, and precision in a three-handed watch with an elegant style. The sale of Zenith to LVMH meant a radical shift in terms of the company's public image and product design. Based on Zenith's technical competence, new CEO Thierry Nataf wanted to position the brand at the leading edge. In addition to extensive sponsorship and marketing operations, this also led to a significant price increase. Watches based on the Zenith El Primero and Elite movements, with classic designs and also a martial design, sometimes extreme-sized, were positioned in the market for both women and men.

In this process emerged such unusual designs as the Defy Xtreme Tourbillon, based on the El Primero, or the Mega Port Royal Open Grande Date and Defy Classic Open series, also driven by this fast-beat movement. The skeletonized balance is the hallmark of these models. Of course, there is still the classic style, in which the Chronomaster, that beautiful calendar watch with moon phase, was enhanced with a flyback complication. The Class Traveller Repetition Minutes features a minute repeater with chronograph, big date, dual time display, alarm and vibrate functions, and dual power reserve indicator. The semi-open dial gives a clear view of the automatic movement balance of the El Primero 4031. These watches certainly keep Zenith in the public eye.

The Zenith commercial and administrative buildings in Le Locle.

12
CHRONOSWISS
MSA
10 15 20 25
DIGITEUR
35 40 45
8
CHRONOSWISS
MSA
20 25 30 35
DIGITEUR
35 40 45

Chronoswiss and Rüdiger Lang—Authentic Feel through the Fascination of Mechanics

In the beginning, watchmaker Rüdiger Lang had only worked for the Swiss watch company Heuer. Then suddenly, people began approaching him, because they knew he was no friend of quartz watches. They asked him to look around Switzerland to see "if there still were a few old mechanical watch pieces," primarily with moon phase. Alfred Rochat, who bought his replacement parts from Lang, still had some—300 altogether, of which Lang acquired 20. He showed them to watch collector Gisbert Brunner, who had been looking for vintage pieces in watch shops—at that time, no one wanted them. He was thrilled, which encouraged Lang in his initiative.

He sold these watches initially still with the inscription Rochat on the dial, yet also already with a transparent display back, to show his customers that they had bought something beautiful. The watches were overhauled accordingly.

In the second year, he said to himself: "If you want to sell watches, you have you come up with a name, a company name." Just how he finally came to the name Chronoswiss, can't be determined any more today. But it was printed from then on the dials, and in 1984 it was registered as a trademark. Since mechanical wristwatches had disappeared from store display windows, something of a wristwatch collecters' movement began to develop. Prior to this, collecting wristwatches was almost frowned upon. Watch enthusiats had collected only pocket watches or precision pendulum clocks.

It was this clientele that Lang made his special target. He mounted good-quality leather straps on his watches and packaged them in an elegant case, imprinted "Alfred Rochat, Montres Compliqués."

Initially, he had a manual-winding watch in his collection, then added an automatic model. Rüdiger Lang also visited other manufacturers in Switzerland, such as the Kelek company, which he knew also made mechanical watches; these included Numa Jeannin or Lemania, Le Phare or Schaffo.

By the 1930s, Eterna was featuring jumping hours and digital seconds and minutes display, mechanically driven. Chronoswiss brought this design back to life for the 21st century.

Chronoswiss head Rüdiger Lang is the great communicator of mechanical watches. He has an outstanding chronograph collection and gladly explains the functions in a model—here a bar lever escapement.

From Kurt Schaffo he bought his first skeleton watches and so brought together an entire range, laid out in an initial catalogue of mechanical watches. That was in 1985—at a time when quartz technology, thanks to mass production in the Far East, was becoming ever-cheaper. Rüdiger Lang kept expanding his collection, step by step. There was even a Kelek repeater. It was not a very attractive watch, so, for the first time Lang commissioned a design genuinely his own from Kelek. Then followed another one and another again, until Lang finally said to himself: "What you're doing now, others can too. They buy watches, put a name on it and sell it to someone else. If you want to set yourself apart from the others, you need to make your own watch." That was in 1988.

This was prior to the big boom, which only started in 1990. Rochat still had vintage Valjoux Caliber 23s and 72s. Limited series were made from these, and shortly after came the Chronoswiss Big Bang: the first Régulateur. This dial design, originally intended for observation watches, was unique in the wristwatch market. The Régulateur is today still Chronoswiss's best selling model. And those purchasing a Régulateur are often first-time buyers at Chronoswiss. The same applies to those who purchase a Kairos automatic, who in any case still doesn't own a model from the Opus or Delphis series. But the next time he purchases a watch, it will almost certainly be one from the Chronoswiss family of models.

From the beginning, the company has striven to always feature or introduce something special in the collection. These include such unusual details as genuine enamel dials. Other manufacturers had and have occasional Special Series with enamel dials, but the Orea became a permanent, established series. An enamel dial requires some technical alterations, because it is twice as thick as a normal metal dial. It is necessary to heighten and lengthen the motion work, canon pinion, and hour wheel and make technical changes, which increase the cost of the production process.

"Since I myself have long been collecting watches," says Lang, "in a Chronoswiss, the details also play a major role. This includes the presentation box and, if possible, even the warranty." The company specially makes its own boxes to match the watch. Moreover, each Chronoswiss buyer receives a certificate of guarantee, hand-signed by the company head. He is not only a chronograph specialist who began work in a chronograph company. He also has an amazing chronograph collection and has written a book on the chronograph.

After Chronoswiss had already made two large-scale chronograph special series, the Kairos with column wheel Calibers 23 and 72, the company presented the Chronoscope, a column wheel chronograph with regulator dial. It is the first column wheel chronograph with regulator dial on the market worldwide. An automatic caliber C.122 Régulateur is ticking in the Chronoscope, named C.125 in the Chronoscope version. Chronoswiss mounts the complex cadrature, or under-dial work, developed in collaboration with the young watchmaker Andreas Strehler, directly on the front side of the bottom plate. To accommodate the components of the complex column wheel mechanism, it is given special milled slots and bores. The three start, stop, and reset functions of the centrally set chronograph seconds are controlled externally via a push-piece integrated in the winding crown and internally via a classic ratchet wheel. This revolves around the shaft of the permanently running second hand at the six. The chronograph wheel is mounted on ball bearings and consists of 38 parts.

The case has a diameter of 38 mm and is composed of 23 parts. It is available, in different designs, in precious metal and gold versions (white, yellow and rose gold) and platinum. The delicate hands of the world's first column wheel chronographs with regulator dial, created in cooperation with the Bielefeld-based Aguilla company, are purple. A sapphire crystal with anti-reflective coating on both sides enhances both hands and dial, which are also available in black and with skeletonized column wheel. The balance oscillations and rotating gilded rotor can be observed through the one-sided anti-reflective coated sapphire crystal.

Rüdiger Lang has a very personal relationship with his watches, his customers, and his sales strategy. As he says: "I have not made watches for dealers, and I have also not made them for myself. The first watch, yes, always for myself, because I have to decide how it works. But I would never want to be in a situation that, in ten years, someone comes to me and says, 'I bought a watch from you ten years ago, which is offered for far less cost today.' Or that the watch is no longer aesthetically pleasing. I am obligated to these people. I also had to save up my money. This is important to me. Therefore, I also said, I have to apply totally different standards. My clientele is not thirty, rather forty-five. The range above and below is ten years. That's 90% of my customers. That means that people have studied brochures; they have made comparisons. For them, the money they spend is important. They will visit a store several times, with the possibility of ultimately buying a Chronoswiss. Someone buys a Chronoswiss, who has selected it very carefully, and knows full well, 'now I want it.' They have been prowling around the store for six months. They always go back in and keep turning it over. 'Do I take it or not?' And then they own the watch. Then it will be a part of it themselves. It's about small but important things. And everything I do for a Chronoswiss is important. I don't do anything special. But I'm doing it consistently. And I always do it for me.

"And I am very, very fussy. I love watches, and so, I make them. And so I sell them as well. I tell people that I like to have my watches. And I use reference points with the watches, not striking, but consistent aesthetic points, and therefore, I am not always making new models. At least, not new cases."

No wonder that a manufacturer who responds so personally and individually to its customers, has a loyal customer base, who are 100% convinced about the watches it makes.

Watch of the Year 2003: the Chronoscope. No other watch of his collection shows more clearly what Rüdiger Lang's roots and intentions are. The first traditional single push-piece chronograph with regulator dial.

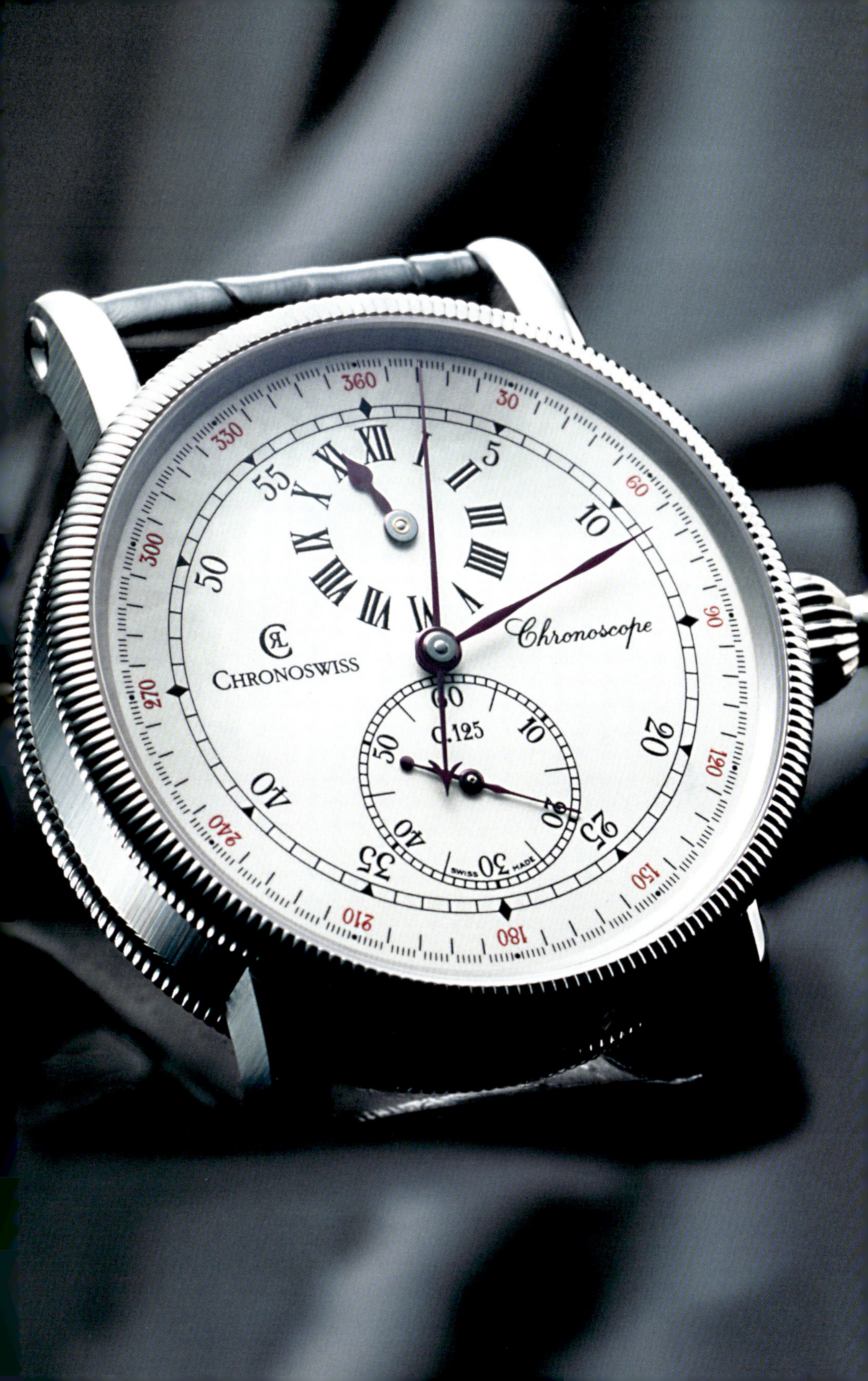
Chronoscope
CHRONOSWISS
C.125
SWISS MADE

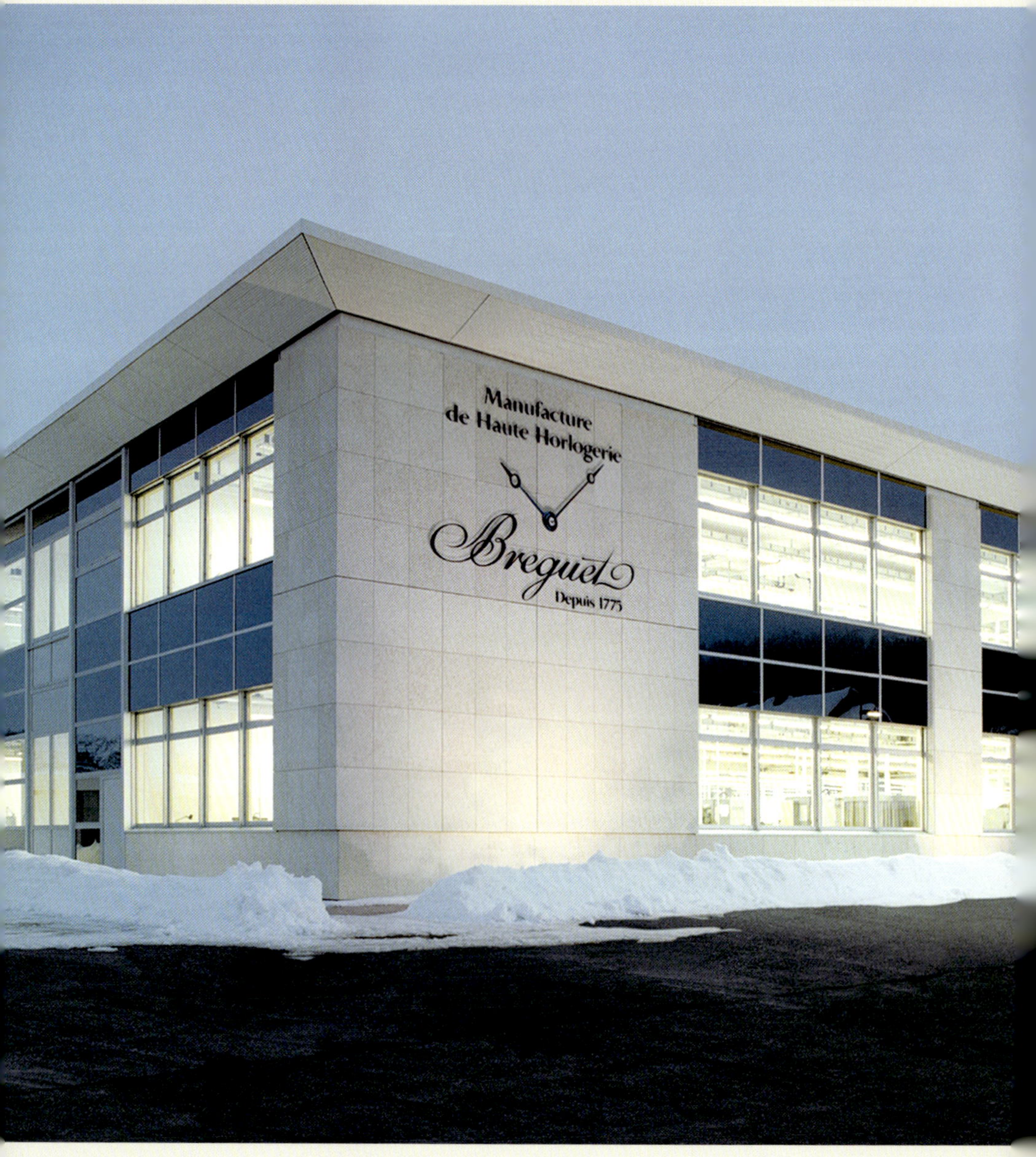
Manufacture
de Haute Horlogerie
Breguet
Depuis 1775

The Swatch Group—Combined Luxury Brands

When the Swiss watch market collapsed into volume segmentation at the beginning of the 1970s, triggered by the chip-controlled digital watch, good advice was expensive. At first, it was the cheap American watches. Next, Japanese makers were soon supplying watches at prices well below those of their Swiss competitors—with serious consequences for the Swiss watch industry. Of the 90,000 employees once working in this industry branch, only half remained. Some 1,000 factories closed down, and, in 1985, those remaining consolidated on the recommendation of corporate reorganizer Nicolas G. Hayek, into the *Schweizerischen Gesellschaft für Mikroelektronik und Uhren* (Swiss Corporation for Microelectronics and Watches), abbreviated SMH. In 1998, SMH became the Swatch Group.

After purchasing numerous watch brands over recent years, the Swatch Group emerged as the largest watch company in the world, which today includes the following watchakers: Breguet (Switzerland) as the flagship, with Blancpain (Switzerland), Jaquet Droz (Switzerland), Leon Hatot, and Omega (both Switzerland) in the top echelon.

In the middle class are such traditional Swiss brands as Rado, Longines, Tissot, Hamilton, Certina, and Mido. The requirements of younger buyers are met by Balmain, Calvin Klein, Swatch, and Flik Flak (all Switzerland). Two German brands, Glashütte Original and Union Glashütte, also belong to the Group. The German Watch Museum in Glashütte can also trace its origins back to a Nicolas G. Hayek foundation.

The real strength of the Swatch Group is the manufacturers of movements like ETA SA in Grenchen, Frédéric Piguet (movement blanks), Lemania (movement blanks), Unitas (movement blanks), and Valjoux (movement blanks). The ETA is the world's largest blank movement manufacturer and produces about 5 million pieces a year. Almost all manufacturers purchase their movements here. With their wide range of mechanical and quartz calibers, the ETA, which emerged from Ebauches SA, is the most important component of the Swatch Group. They are joined by suppliers for mechanical watches such as the mainspring maker Nivarox and even those for quartz watches, such as the chipmaker EM, Micro Crystal for quartz production, Oscilloquartz (crystal oscillators), and battery manufacturer

Breguet, flagship brand of the Swatch Group and declared favorite child of senior director Nicolas G. Hayek.

Always good for exceptional complications. Beyond power reserve indicator and perpetual calendar, the Äquation features an equation of time display.

Renata. With Georges Ruedin SA, they have, in addition, a case-maker, and, with Universo, a watch hand manufacturer in the network. Other companies are Swiss Timing, which deals in effective advertising involving sports timekeeping, and LASAG for laser measurement technology and SOKYMAT for automotive transponder technology.

What made the turn-around possible, in addition to concentrating the remaining companies, was a watch that marks the Group name still today: the Swatch. By the mid-1990s, more than 150 million had been made. Together with the physician Ernst Thomke, Hayek had configured a watch, comprised of about 54 parts, which is assembled as fully automatic watch and could be sold profitably at a price of approximately $125. The collection is revised twice a year. Initially it included quartz watches with analog display; from 1991 on, mechanical movements with ETA Automatic caliber 2840 became available in Swatch models.

Skillful product placement and employing internationally known artists to design the timepieces, have made these watches cult objects. An international collectors' scene developed; some have at times paid up to $40,000 for exceptional models. The prerequisite is that a watch is in brand new condition. In 1993, there was even a platinum Swatch model (the Trésor Magique) offered for about $3,500.

"A luxury watch also for sportswear"—the message of the Marine model with rubber strap and big date in stainless steel.

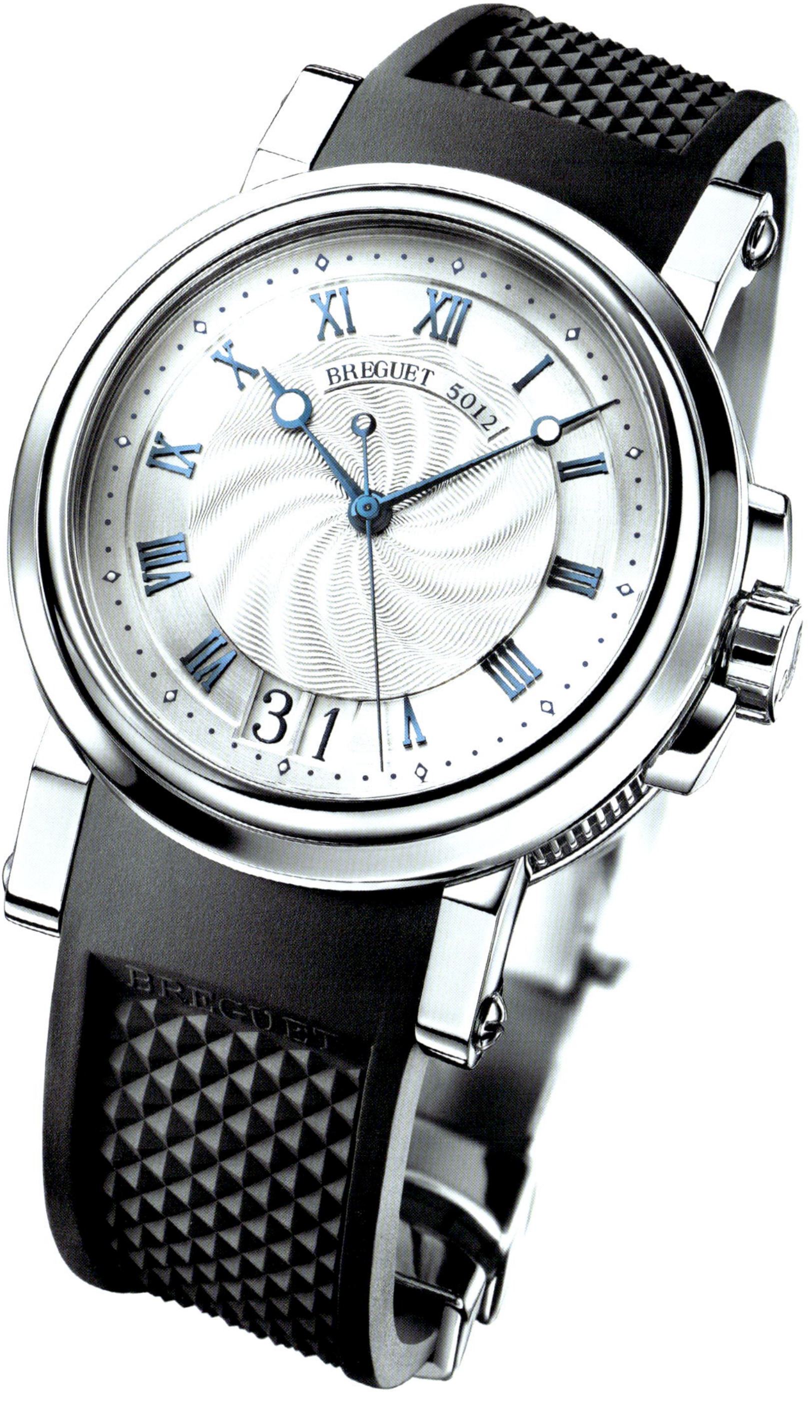
BREGUET 5012
31
BREGUET

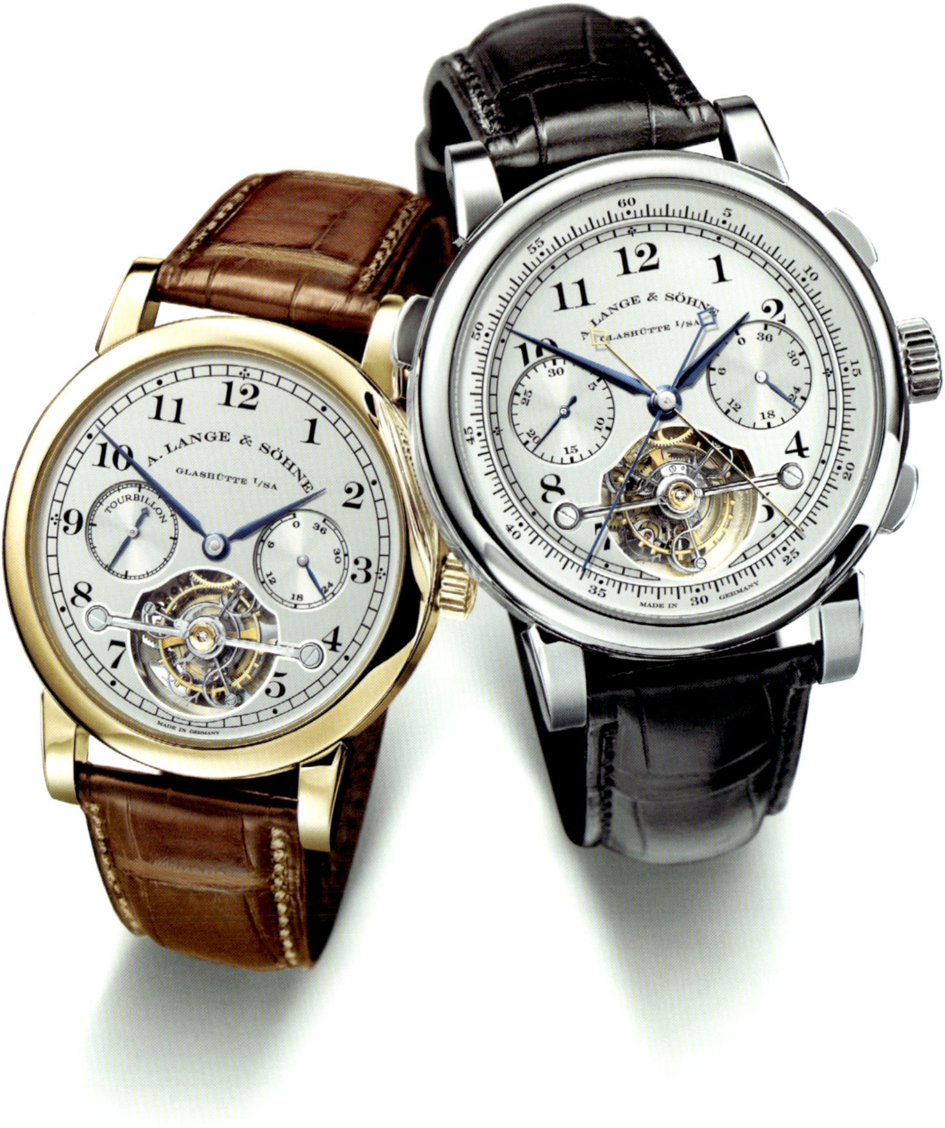
A. LANGE & SÖHNE
GLASHÜTTE I/SA
TOURBILLON

Wearing Time in Fine Style—A. Lange & Söhne

Salzburg in summer. An elegant Mercedes-Benz W111 series coupe drives up, and an elderly gray-haired man emerges with a springy step. You drink a cup of coffee together. The watch enthusiast notices that his companion is wearing an extraordinary timepiece on his wrist. An A. Lange & Sohne 1 with blued—not, as usual, golden—hands. When asked about it, he smiles mischievously and explains to the stunned party opposite, that this is his photo watch, which, as a great-grandson of company founder Ferdinand Adolph Lange, he wears as the representative of A. Lange & Söhne to ensure better visibility in photographs.

Even if we have heard nothing of them for a long time, the great brands do not really ever disappear from the market. Often they shine brighter in retrospect than when they were active. With this phenomenon is linked, for many admirers, the hope that they will return and their old splendor will shine anew. In such cases, it sometimes requires only an external impulse to set the wheels in motion again, and, at one stroke, Maybach sedans will be on the road and A. Lange & Söhne watches will begin ticking again.

While for Maybach, it was the celebration of Wilhelm Maybach's 150th birthday which brought the name back into the awareness of the Daimler-Benz company. In the case of Lange watches, it was German reunification—there where its *genius loci* (protective spirit) lingered, in Glashütte—that made a renewed start possible.

But wait! We will tell the story of Walter Lang and the watches with the same name, from the beginning. The year was 1845, when his great-grandfather, supported by a state loan granted in the amount of 5,580 thalers by the Royal Saxon Interior Minister, began to build up a watchmaking operation in Glashütte. On December 7 of that year, the factory was officially inaugurated, and from then on, highest-quality fine pocket watches were made in this small Saxon town. Tourbillons, perpetual calendars, chronographs, Grande Complications, and

Two A. Lange & Söhne tourbillons. The Tourbograph (right) is one of the most complicated watches the firm offers. It is powered by fusée-and-chain. Next to it, the Sachsen Tourbillon, known in the firm as the Pour le Mérite (the highest Prussian order of merit).

even a self-winding pocket watch left the factory in order to beat or display the time for the lucky few. The Lange Grande Complication made in 1908 had a value roughly equivalent to that two houses. For about 4,930 gold marks, the delighted owner received a clock with perpetual calendar, minute repeater, 4/4 self-striking chime, and a rattrapante chronograph.

With great skill, the Lange family also directed the company through difficult times, and Walter Lange, born in 1924, completed his training as a master watchmaker in Karlstein and Glashütte. But then the Second World War broke out. When Walter Lange, who had been drafted into the army in 1942, returned to Glashütte in 1945 gravely wounded, he had to look on as, just a few hours before the war ended, the main building of the factory was almost completely destroyed by bombs. The parental operation, nationalized in 1948, joined a combine in 1951, and Walter Lange, who refused to join the FDGB (*Freier Deutscher Gewerkschaftsbund*, or Free German Trade Union Confederation), was conscripted to work in the uranium mines. It was time to pack his bags and leave for the West. He

A new administration and exhibition building was built next to the building internally designated "Lange 1."

and his young family found their new home in the watch and jewelry city of Pforzheim. Walter Lange was now working in the jewelry industry, including work as a sales representative for the Wellendorf company. It is at this time that the first delicate contacts with IWC Schaffhausen began to develop, which even then had in hand producing Lange pocket watches as a joint project; this failed due to the burgeoning quartz watch market.

When what no one had thought possible happened, and the Berlin wall fell in 1989, the idea of a new beginning in the company's ancestral home emerged, in cooperation with IWC managers Günther Blumlein and Hannes Pantli. The rebirth of A. Lange & Söhne would cost, according to Blümlein's estimate, some 500,000 DM. In the end, it cost more than 20 million to attain the ambitious goal of making the highest quality watches in the world.

On the day, exactly 145 years after the founding of the company by Walter Lange's great-grandfather, on Dec. 7, 1990, the firm was entered in the Commercial Registry. Many difficulties had to be overcome, until, on Oct. 24, 1994, the Lange 1 models, the Saxonia, Arkade and Tourbillon Pour le Mérite were presented to an admiring public. The patented Big Date represents a special feature for these watches. To ensure that there were as many pictures as possible in the press the next day, the date on all the watches was set for the 25th. Marketing operations also began with a bang. A two-page color ad appeared in the *Frankfurter Allgemeine Zeitung* with following contents: "The economy in the East is suddenly beginning to tick in a very different way: A. Lange & Söhne is back—the legend has become a watch once more."

Reactions were overwhelming. The watches from Saxony took the hearts of aficionados by storm. In 1995, Walter Lange became an honorary citizen of Glashütte and three years later received the *Verdienstorden des Freistaates Sachsen* (Order of Merit of the Free State of Saxony.) Meanwhile, in Glashütte watches were once again being made to make the heart of any connoisseur and enthusiast beat faster. Each watch series has its own movement. The three-quarter bottom plate made of untreated German silver and hand-engraved balance cock are characteristic of Lange watch movements. There are visual splashes of color in the blued screws and ruby-red jewels.

The collection was continuously expanded to include further complications. In 1999, a chronograph appeared with jumping minute counter; in 2001, a perpetual calendar; and in 2004, a Double Chronograph, which has a split-seconds mechanism for the minute counter and thus allows comparative measurements for a period up to 30 minutes. The latest model is the Lange 1 Time Zone. This watch can show the local time for all the major cities of the world, displayed on a ring which can be adjusted by push-piece, set in a small sub-dial at the five. Besides its day or night indicator for the time zone display, this manually winding watch has another small complication, a power reserve indicator, and the famous Lange Big Date.

More fascinating complications followed. These include the Double Split, a manual-winding split-seconds chronograph, where this rattrapante mechanism makes it possible for the first time to record times differing by up to 30 minutes. Anyone admiring the 465 movement parts through the sapphire crystal back, will have to ungrudgingly attest that A. Lange & Söhne has reached the highest level of movement refinement. The inventor of the big date is still featuring it in all versions of its watches, the manually winding chrongraph, moon phase watch, and of course, the perpetual calendar. The latest delicacy is a tourbillon in Cabaret case; the tourbillon carriage can be stopped by pulling out the crown.

What hardly any other manufacturer does: At A. Lange & Söhne, every case series has its own movement. This alone ensures a completely exclusive quality. All new designs are evaluated by Walter Lange himself. Do technology and design accord with the essence of A. Lange & Sohne? Only when he gives his nod, will a model go into series production. But Lange is not only a connoisseur of watches, but also of automobiles. For him, the most beautiful classic car is the Mercedes W111 250 SE coupe, which was manufactured between 1965 and 1967. Its 150 hp fuel-injection

Assembling the Big Date disc for a perpetual calendar with moon phase display.

engine already then generated a top speed of 190 km/h and accelerating from 0 to 100 km/h in 11.8 seconds. The single-joint swing axle with hydro-pneumatic compensating springs and self-leveling suspension ensured ride comfort and secure road holding. The dashboard is lavishly crafted from a combination of wood and leather.

The car, delivered on Nov. 25, 1966, cost back then a full 27,493 DM (about $7,000). This price brought the new owner green leather seats, power steering, an electric steel sunroof, and a Becker Grand Prix radio. The ivory-color steering wheel delights both sight and feel; it still gives the

Several hundred thousand euros worth of watches on the watch winder in the testing lab. Their precision has to be demonstrated before they are shipped out.

car that extra touch today. Whenever his tight schedule allows it, Walter Lange enjoys a ride in his classic Mercedes.

In addition to his many commitments involving "his" watches, Lange published his memoirs in 2004. Its title can also be confidently be taken as a summing up of his lifelong work for the cause of A. Lange & Söhne: *When Time Came Home.*

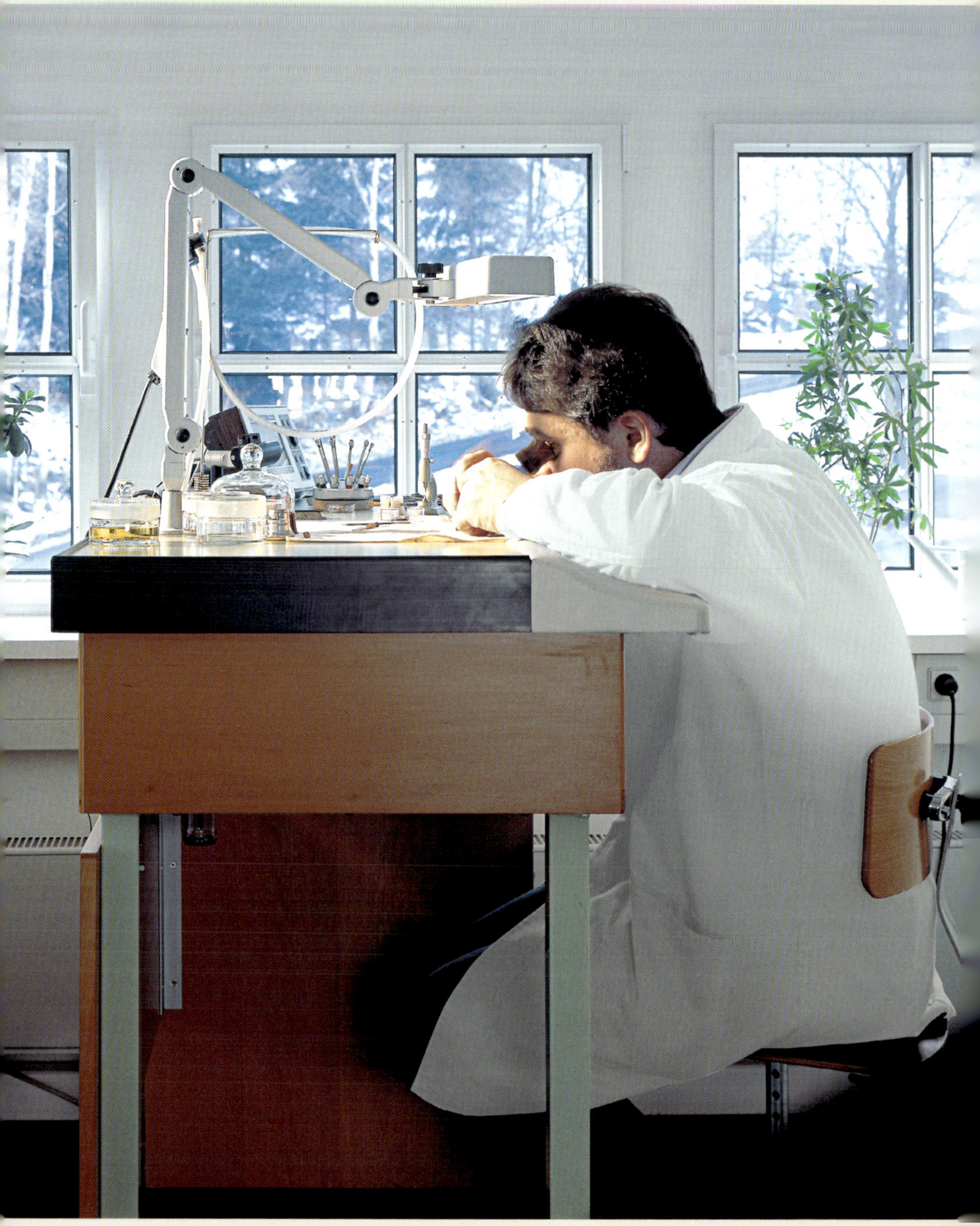

The Fine Craft Workshop in the Müglitztal: How Are A. Lange & Söhne Watches Created?

After crossing the Müglitztal, just after the entering the town of Glashütte, you will see the Lange & Söhne factory building on the left. The number of employees has grown from about 270 to 500. Half of them are watchmakers, but this workshop is still making only 5,500 watches a year. In comparison, Rolex makes approximately 1 million watches. But the people of Saxony want, alongside quality—something, beyond question, that the Swiss also have to offer—primarily to enhance the complexity of their products, that is, to create sophisticated and unprecedented complications, such as a tourbillon with a cage that can be halted for precise adjustment of the time, or the Tourbograph, which combines a split-seconds chronograph with a tourbillon.

In these fascinating watches, it is primarily their internal beauty that distinguishes a Lange & Sohne watch from most of their competitors. The untreated German silver for the movement parts forms the basis for a masterful staging for skillful handcrafting.

Lange brought back to life many elements that were used in making historical pocket watches. These include the three-quarter plate or screw-mounted gold chatons. The various finishes and perlage are given to areas which the wearer cannot see—actually, an attractive anachronism. Originally, this type of finish was used to bind the dust that penetrated the watch case, and thus keep it from getting caught in the train and negatively affecting the watch's timekeeping performance.

However, they did not just carry on using traditional features, but developed innovations to newly interpret the mechanical watch, such as the patented Big Date or the Zero-reset to set the hands. When the crown is pulled, this jumps the second hand to zero and it pauses there until the crown is pressed down again. This makes it possible to set the time to the very second. All watch parts are decorated and finished by specialists in laborious handwork.

Therefore it's no wonder that you will have to invest at least $17,000 to wear a

"Time loves nothing that happens without it." Lange watchmakers are also committed to this principle. Each watch is assembled with all due thoroughness.

new Lange watch on your wrist. You have to experience how much meticulousness and dedication are invested in making these watches, and Walter Lange gets to the point: "Anyone holding a timepiece from A. Lange & Söhne in their hands, immediately recognizes that they have something special in front of them. But to really understand what constitutes its unique quality, you have to see how a Lange watch comes into being in its home town, in Glashütte in Saxony. Because only those who visit the Lange workshop and are allowed to watch over the shoulders of the master watchmakers at their concentrated and highly intricate work, can realize how much craftsmanship, precision, and, above all, passion are contained in a watch with the signature of A. Lange & Söhne."

An attentive viewer will also be struck by the fact that each Lange model houses its own individual movement, which was designed specifically for the corresponding watch. This applies not only to the Arkade- and Cabaret-shaped movements, which are each distinct, but also to their round brothers and sisters. No other manufacturer in the world has carried out its work in this area with such iron determination as Lange. In a Lange watch, you could never imagine discovering a round movement in a rectangular case. They also exclusively use precious metal cases, of platinum, white gold, rose gold, and yellow gold. What began in 1994 with the Lange 1, led, by 2008, to a considerable family of calibers: model Lange 1, Caliber L901.0; model Grand Lange 1, Caliber L901.2; model Lange 1 Moon Phase, Caliber L901.5; model Lange 1 Time Zone, Caliber L031.1; model Saxonia, Caliber L941.1; model Saxonia Automatic, Caliber L921.4; model Grand Saxonia Automatic, Caliber L921.2; model Richard Lange Caliber L041.2; model Lange 31 Caliber L034.1; model Langematik Perpetual, Caliber L922.1; model Datograph, Caliber L951.1; model Datograph Perpetual, Caliber L952.1; model Lange Double Split, Caliber L001.1; model Cabaret, Caliber L931.3; model Cabaret Moonphase, Caliber L931.5; model Cabaret Tourbillon, Caliber L042.1; model Arcade, Caliber L911.4.

In the special series, which are now mostly sold out, there are other calibers: model Tourbillon Pour le Mérite, Caliber L902.0; model Lange 1A, Caliber L901.1; model 1815 Moon phase, Caliber L943.1; model Lange 1 Tourbillon , Caliber L961.1; model Jubilee Langematik, Caliber L921.7; model Grand Lange 1 Luna Mundi , Caliber L901.7 + 8; model Tourbograph Pour le Mérite, Caliber L903.0; model 1815 Calendar Week, Caliber L045.1. The Chronograph 1815, where the Big Date was dispensed with, is no longer in the production program.

Lange watches are not custom made, yet it is legitimate to compare them throughout with any unique piece. The balance cock of every Lange watch is decorated by a master engraver in freehand engraving, making it unique. Anyone who is able to show his own Lange watch to one of these engravers, will learn exactly who among their artist colleagues did the work on the balance cock of his

A lovely scene: Pocket watch on the table, wristwatch on the wrist of the watchmaker working on an escapement model.

watch. With any luck, you can actually obtain a picture of the engraving autographed by the engraver responsible. All parts of each watch are exactly matched to each other. As a result, you cannot automatically use the balance of one Lange watch in another watch, even if it involves the same model.

The largest and heaviest part of each Lange movement is the movement plate, which weighs 5.42 grams. In the Lange 1, it has a diameter of 30.40 millimeters and a height of 1.40 millimeters. In contrast, the smallest and lightest watch movement part is the thinnest of the washers for the balance

145 years after Adolf Lange's pioneering feat, Walter Lange founded this old and new company in Glashütte on Dec. 7, 1990. Time had come back home

wheel screw. It has a diameter of 0.4 millimeters, a height of 0.01 millimeters, and a weight of 8 micrograms (0.000008 g). In other words: it takes 125,000 tiny washers to make a full gram. The movement of the Datograph Perpetual comprises 556 parts, closely followed by the Langematik Perpetual, with 478, and the Lange Double Split and Pour le Mérite Tourbograph, each with 465 separate parts. If its chain were separated into its component parts, the Tourbograph would even have 1,097 components. As to the Tourbograph—it is currently the most complicated and expensive—costing about $470,000—Lange watch. Its additional title *"Pour le Mérite,"* literally translated "for the merit," was an order of merit created by Friedrich Wilhelm IV and Alexander von

Humboldt in 1842, awarded for academic, but also for military achievments, and was used both in the past and present to designate Lange's top-quality products.

The Tourbograph Pour le Mérite is a watch without precedent. It is the first minute tourbillon in wristwatch form with fusée and chain transmission, which had an additional split-seconds chronograph. The fusée and chain mechanism serves the same purpose as the tourbillon: to positively influence the rate of the watch. The different degrees of torque at the beginning and end of the watch-winding process, are equalized by running the chain over a cone-shaped fuse, so that the power is released uniformly, independent of the stage of winding. How does this work? The mainspring barrel and fusée are connected by fine chain, of over 600 individual pieces. When fully wound, meaning at great tension, the chain pulls on the smaller circumference of the fusée, i.e., as on a smaller lever; as it is unwound, decreasing tension, then it pulls on the larger circumference of the fuse, or on a larger lever. This technology has been used since 1880 in A. Lange & Söhne's pocket chronometers, as well as the Tourbillon presented in 1994 in the first collection, which was also was called Pour le Mérite.

Another watch with chain transmission was introduced in 2009 as the Richard Lange, which now also has the additional title Pour le Mérite. In terms of what level of precision can be achieved in a watch movement, transmission by chain and fusée is even superior to the tourbillon. G.H. Baillie wrote on this in his standard work *Watchmakers and Clockmakers of the World*, that "… probably no problem of mechanics was ever solved so simply and perfectly!" The Richard Lange Pour le Mérite, which consists of an incredible 636 individual parts, externally creates the modest and sober impression of a simple three-hand watch; this predestines it to be the understated watch *par excellence*. The exclusive Caliber L044.1 has a constant-force escapement that works at 21,600 vph. The power reserve is 36 hours when fully wound. Lange makes the mainsprings themselves and tailors them to the movement's requirements. The inertia torque of the balance, balanced with 18 ground screws and four adjusting screws of solid gold, again corresponds exactly to the drive torque of the fusée. Once more, A. Lange & Sohne is presenting not only a highly accurate watch, but also a special piece of craftsmanship.

A unique movement is ticking in this new, extraordinary watch, as in all the other Lange models; these are created in Lange's design department with respect to their own product history. Almost all parts of these exclusive watch movements, which comprise in total some several thousand different pieces, are manufactured in the workshop—including plates, bridges, levers, springs, wheels, and pinions. Each part gets an elaborate surface finish. Even all the areas that remain invisible to the owner in the fully assembled movement are decorated. As a result, Lange's specialists take two days alone just to decorate the delicate tourbillon bridge, which mounts the tourbillion on the side of the dial. For manual perlage of the bridges and plates, they use up to 14 special tools for each different perlage diameter. The tools and mounts needed for grinding and polishing are produced in Lange's own tool shop. The balance cock is given a freehand engraving, thus making any Lange watch a unique piece. The movement is assembled by Lange watchmakers, carefully adjusted in five positions, and then disassembled again. Only then are the individual parts given their final finish; they are cleaned again and—only now using a finely blued original screw—re-assembled to make a perfect watch movement. Before the watch is allowed to leave the factory, it will be given several weeks of inspections.

The company's future goal is to increase the in-house creation of value. To realize this, in 2003, the new technology and development center was built, the manufacturer's fifth building. Here they deal with the theoretical fundamentals and technological processes required to produce their own balance wheels, the heart of any mechanical watch. From drawing a wire to a thickness of 0.05 millimeters, to rolling, coiling, and annealing, and up to coiling: all the work stages to make highest-quality springs for the various Lange movements are now carried out in-house. The first to be utilized was a balance wheel developed and manufactured by Lange themselves, in the Double Split chronograph introduced in April 2004.

Among rectangular watches, which have been available since 1994 and 1997 respectively as the Arkade and Cabaret, an interesting new design was created in 2008 with the Cabaret Tourbillon, something previously unknown to the world of watches. While the tourbillon had been patented already by 1801, the stop-seconds, which lets you to set the time so precisely, was never available before. For the first time, Lange designers had succeeded in taming the whirlwind—even briefly. Added to this system for ensuring the watch's precision, now you cam set it exactly to the second. From the beginning, they dispensed with the option of mechanically stopping the entire tourbillon cage from inside the movement. This rather simple solution would cause the balance to stop pivoting, and eventually run out of energy. Not a technically satisfactory solution, because starting it again would require some external thrust. To preserve the potential energy of the balance spring at the moment of braking, requires direct, lag-free braking of the balance rotating in the cage. The only question was how to stop the oscillating balance of a tourbillon escapement in a rotating cage. The answer: pulling the crown, which triggers a complex leverage movement, that causes a stop lever with two V-shaped, curved spring arms to touch the outside of the balance wheel, stopping the balance instantly. This process is complicated by the fact that the V-shaped braking spring could also land on one of the tourbillon cage's three pillars.

For this reason, the fine, two-armed, steel mobile stop-spring is mounted on a rotation point of the brake lever. That is, one spring arm lies against the cage pillar, while the other is lowered to the balance rim, and halts it just as reliably as if both spring arms had met the balance rim. The asymmetrically curved shape of the two spring ends was determined through a long series of tests. They are precisely formed so that they exert optimum pressure on the brake spring, in all balance positions. In addition, the ends of brake spring are curved, so they cannot get caught as the balance is stopped and released.

All in all, this again was a typical Lange solution, which no one had ever conceived of before.

A. Lange & Sohne, founded in 1994 as part of Mannesmann Group-member LMH, with the assistance of its sister companies IWC and Jaeger-LeCoultre, is today one of the world's most exclusive watch brands.

In 2001, the watch conglomerate was acquired by the Swiss Richemont Group, which brings together some of the leading manufacturers of fine watches, including, besides Lange, IWC, Jaeger-LeCoultre, Baume & Mercier, Piaget, Vacheron Constantin, Officine Panerai, and Roger Dubuis.

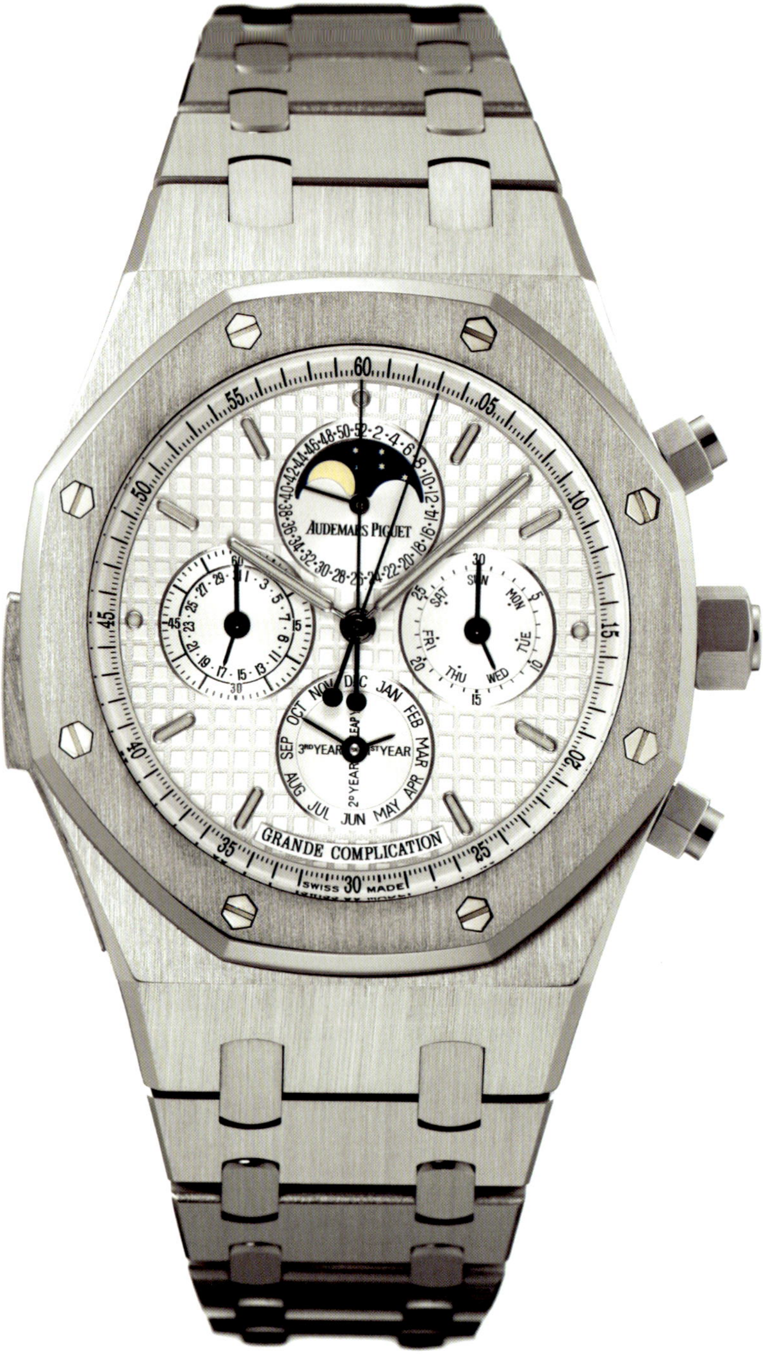
AUDEMARS PIGUET
GRANDE COMPLICATION
SWISS MADE

Audemars Piguet—The Myth of a Great Brand

Audemars Piguet is a major watch brand that is rich in traditions. Founded in 1865, it is still today owned by the founding family. Part of the *Haute Horlogerie* through every decade, this company is a true master of mechanical watch complications. AP had already made the first wristwatch with minute repeater by 1892, and a year earlier, it had presented the only 18 mm wide movement. It is always surprising its competition and customers—as in 1924 with a wristwatch with jumping hour display, and a year later with the thinnest pocket watch in the world, with a movement height of 1.32 mm. In the Royal Oak, AP presented a luxury sports watch in steel; the automatic version had a winding rotor of 18k gold.

Each decade features an exceptional piece. In 1989, this was the thinnest manually wound movement with perpetual calendar; in 1998, it was a Grande Sonnerie minute repeater with carillon. Some 450 employees, including about 125 in production, make just 18,000 watches a year. This results in an annual sales of over CHF 200 million and rising.

It is only legitimate for such a company to also evoke the power of tradition with its nomenclature. Jules Audemars, a company founder, today provides the name for a special series of watches, most with complications, which are beyond the reach of average income earners. A tourbillon with chronograph, a minute repeater both as Grande Sonnerie carillon and carillon with tourbillon, a perpetual calendar with world time display, and a perpetual calendar with sunrise and sunset and equation of time. Altogether, watches that at times cost significantly more than $65,000, and therefore beyond most peoples' dreams. However, there is one complication in this series still just affordable for the watch aficionado and yet one that stands out from a normal three-hand watch: the chronograph.

In the Jules Audemars range chronograph, Audemars Piguet offers a watch that draws its charm from the right blend of tradition and luxury. This rather simple watch

A Grande Complication with the look and feel of a sporty steel watch, and with a quality only connoisseurs will perceive: minute repeater, perpetual calendar with moon phase, and rattrapante chronograph for everyday sportswear.

No company is more authentically or longer family-owned than Audemars Piguet, next to Patek Philippe the greatest Swiss luxury watch brand. One can only wish both their continued independence.

is made of 18 carat white or rose gold or steel, with a fold-over clasp of the same material. It is the least expensive alternative if you want to purchase an Audemars Piguet watch with complication, and therefore worth a closer look.

At first glance, these are rather unassuming and diminutive watches, which reveal their true class only at the second glance. Weight, including bracelet and clasp: 95 grams. With a case diameter of 39 mm, it is anything but small. Nevertheless, the slight curvature of the case, satin finished on its rim, in combination with the polished lugs and similarly finished bezel, make the watch appear smaller than it actually is; the height of only 11.1 mm also contributes to this. On the case back, secured with five screws, the combination of matt finish and polished rims is continued. Crown and push-pieces are not screwed down, although the knurling at the edging by the push-pieces gives this impression. The watch is nevertheless waterproof to 20 meters. The chronograph comes with a black crocodile leather strap and fold-over clasp. This is made as a stylized Audemars Piguet monogram and is found on all Jules Audemars models, as well as the models of the Edouard Piguet series.

The sapphire crystal is mounted exactly and without any overlap. The baton hands for hours and minutes are of white gold, as are the wedge-shaped indexes and Arabic numerals. The company logo is at the twelve. The dial is designed in blue, but, depending on the light, shimmers between blue and deep black. The hands for the chronograph and totalizer are painted white, like that for the small seconds. With the small seconds at the three and the 30-minute totalizer at the nine, the dial corresponds to classic

The headquarters in Le Brassus, home of Audemars Piguet.

chronograph design, with the also white-printed minute track on the chapter ring. The fine subdivision on the rim of the dial by four division marks is not entirely accurately designed to measure the eighth of a second.

Reference 25859 BA is equipped with the AP Caliber 2226/2841, that also serves in the Royal Oak Offshore, although there with date and hour totalizer. Not only the caliber designation, but also location of crown and push-pieces clearly indicate that this is a modular caliber. The modular chronograph used by AP was originally designed for quartz movements, and therefore had to make do with extremely low power, which now works positively on the rate values of the automatic movement when the chronograph is engaged. The basic movement, a true in-house movement, has a height of 6.15 mm and a diameter of 26 mm (11½ lignes). The rate is 28,800 vph (4 Hz), and the 54-jewel movement has a power reserve of 40 hours. The bi-directional winding rotor has a weight of 21k gold. If the crown is pulled out, the second hand stops, so the watch can be set accurately to the second. The watch's precision, at one second plus or minus, is outstanding, and surpassed every chronometer standard by far. This might console potentially interested buyers, since the modular design without column wheel doesn't quite match the idea you might have of *Haute Horlogerie*. The movement is finished with Geneva stripes, which quite nicely match the gold segment on the end of the rotor: a true in-house movement with an appropriate charisma. As fascinating an effect as the watch creates with its discreet appearance, and however perfect the rate values were, the lack or a pedigreed chronograph movement is still a small drawback. In this price range, you will encounter tough competition, even within the same house, where the Royal Oak model uses a Piguet 1185 re-engineered by AP(Audemars Piguet 2385). Anyone, however, who is looking for a fine line and a subtle look in a chronograph, would do well with this watch, and will also be rewarded with fantastic rate values.

There are more collections from AP, including the Royal Oak, Millenary, Canapé, the Promesse, and the Royal Oak Offshore. Most of the available complications are offered in all watch ranges, as they are in the oval Millenary, which is available both as a Dual Time watch and a chronograph and three-handed watch. Highlights include, beyond the complicated pieces such as the Metropolis with world time display and perpetual calendar, timepieces like the Concept Watch. When you catch sight of this watch, with a case made of the super alloy used in the aerospace, Alacrite 602, titanium bezel and open-work movement with tourbillon, dynamograph (torque indicator) and the display for the still-present barrel rotations, you will involuntarily ask yourself whether all watches would be made today in just such a design today, if the quartz watch had never existed. On the Concept Watch, winding, setting, and neutral are activated by a separate push-piece next to the crown, which there is no need to pull out.

Anyone who can call an Audemars Piguet their own, may also be sure that their watch is also part of the company dispatch books and thus company history. These books have been kept meticulously since about 1882, and make every purchase traceable. Each watch is part of the House of AP's *savoir-faire*; for watch connoisseurs, *savoir-vivre*.

The Jules Audemars Tourbillon minute repeater. This rose gold watch also has a chronograph complication and is comparably priced to the range of luxury sports cars. Next to it, a platinum version without chronograph, which exhudes rather more understatement. Both are manual winding models.

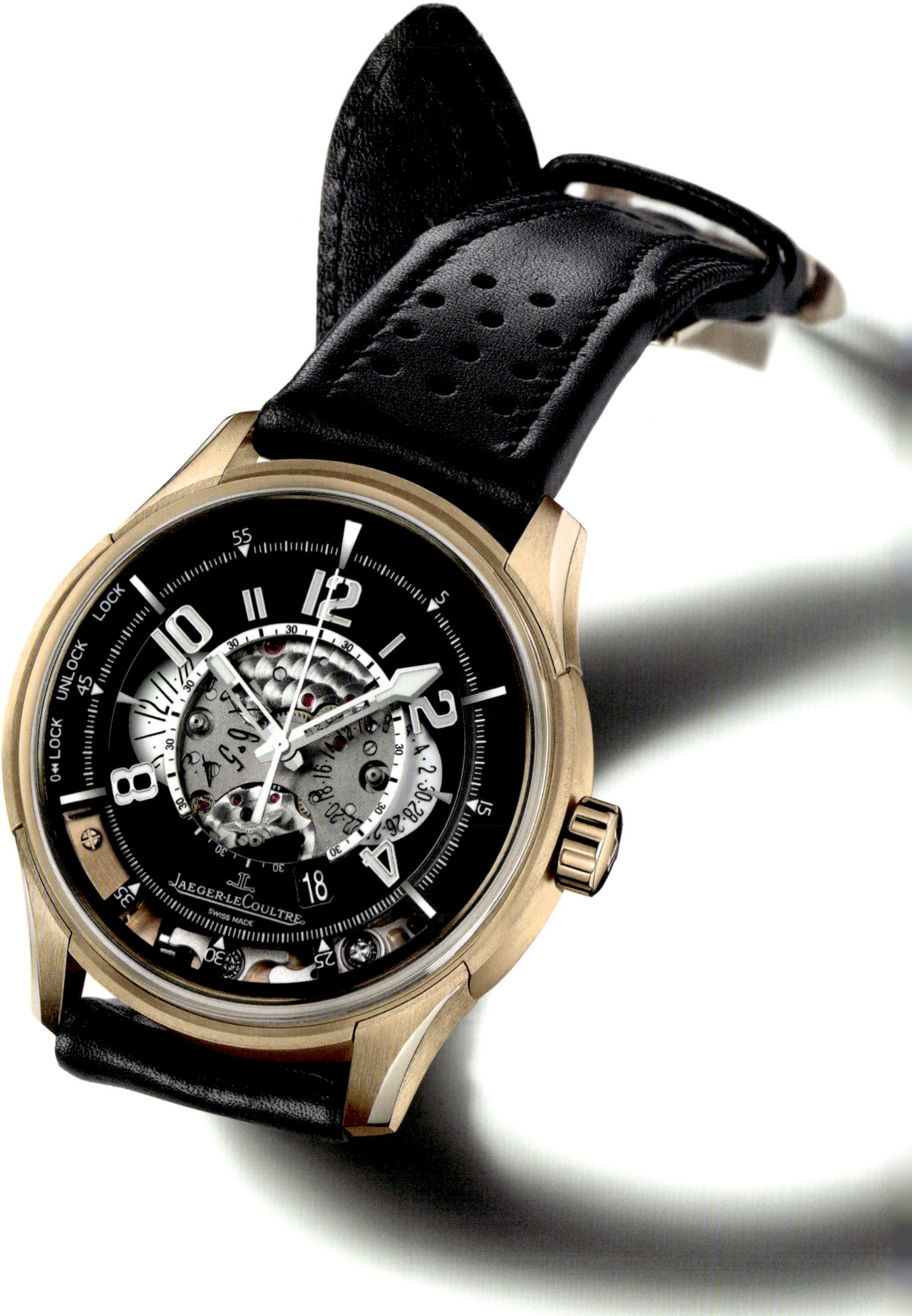
LOCK
UNLOCK
LOCK
JAEGER-LECOULTRE
SWISS MADE

Jaeger-LeCoultre—Fine Workshop and Master of Complications. The Movement Specialist from the Vallée de Joux

On the shore of Lac de Joux lies the village of Le Sentier, home to one of the most famous watchmakers: Jaeger-LeCoultre. Directly opposite the factory main entrance is a monument to the man who launched the company in 1833. Antoine LeCoultre was only 30 years old when he opened a workshop to produce pinions and small gear wheels for watchmakers. When he died in 1881, he had by then fullfilled his life's dream. He was the biggest watch manufacturer in the valley of the watches. Several hundred people worked in his company and also manufactured, besides watches, the machine tools necessary to make them.

Antoine LeCoultre came from a Huguenot family, which, like many others of their faith, had had to leave France. He learned to work with metals in his father's forge. He was a gifted inventor. In 1844, LeCoultre invented the millionometer, a measuring device which could measure one millionth of a meter, one micron. The precision made possible by this device, of course, had a positive effect on watch production.

By 1870, LeCoultre was employing some 200 people. His son, Elie LeCoultre, took over the direction of the company and expanded industrial production. Together with his brother Benjamin, he made the company the prime address for complicated pocket watches with calendars, repeaters, and chronographs. Elie LeCoultre's son Jacques-David, who joined the company in 1899, successfully carried company tradition forward. In 1903, he initated business relations with the Paris-based Alsacian chronometer maker Edmont Jaeger. Jacques-David had to bicycle to the neighboring village to make the critical telephone call, because at this time there was still no phone in Le Sentier. His father remonstrated with him about the expensive telephone calls, even though this was the best investment for the company's future.

Edmont Jaeger's experiences in the world of luxury watches opened totally new perspectives. By 1879, Jaeger had settled in Paris with his own watchmaking workshop,

A striking feature of the Jaeger-LeCoultre Amvox 2 chronograph: it has no push-piece. The start-stop function is activated by pressing the watch crystal. The impulse is transmitted by lever train to the pusher axis.

Antoine LeCoultre (1803-1881), the brilliant creator of the company.

and later became official supplier to the French Navy. In 1907, Jaeger LeCoultre created the long wished-for thin pocket watch movement, which, until today, is the thinnest pocket watch movement with a crown winding mechanism: Caliber 145, with a diameter of 39.54 mm and a height of only 1.38 mm. Jacques-David, who had taken over management in 1906, developed their first wristwatches in the 1920s, in close cooperation with Edmont Jaeger, who died in 1922.

The time was rich in innovations. The Duoplan was a small rectangular movement, that fit well in the rectangular Art Deco style watches popular at the time.

With the Caliber 101, LeCoultre produced the smallest mechanical movement up till today, with dimensions of 11x13.5 mm and a height of 1.5 mm. This movement was ticking in the diamond watch worn by Queen Elizabeth II for her coronation. In 1928, using the Caliber 134, LeCoultre manufactured one of the first wristwatches with eight-day power and alarm function. In the same year, the Atmos, a table clock that draws its energy from the fluctuations of air temperature, made its premiere. It is still made today, as is the Reverso, which was developed in 1931 at the request of polo-playing British officers. Players were repeatedly shattering their watch crystals when playing this rough sport. The reversible case, which allowed the wearer to turn the watch crystal downwards, was a simple solution for the problem. It has gone down in watchmaking history as an Art Deco design classic, and still finds aficionados today. Jaeger and LeCoultre merged in 1937 to create the current Jaeger-LeCoultre trademark. Jacques-David, who died in 1948, was the last of the family to direct the company.

In the 1950s and 1960s, models such as the Geomatic, Memovox, and Futurematic shaped the character of the brand. The Geomatic is a watch with special protection against magnetic influences; the Memovox is the first automatic winding alarm watch, and the Futurematic, with automatic Caliber 497, eliminated the crown. Jaeger-LeCoultre was also involved in the development of the Beta 2, a quartz movement.

Almost unnoticed at the road's edge, the Jaeger-LeCoultre administration and production building in Le Sentier.

Like many others, the company held that a watch's accuracy is its highest precept, and banked on a technology that would prove disastrous for the entire watch industry. Jaeger-LeCoultre was thus not immune to the quartz shock, but they had never lost faith in the mechanical watch, and, from 1980 on, things slowly improved. They had developed an ample number of ideas during the crisis period, although there were many in their own ranks who saw no chance of a future for the mechanical watch.

The fast beat Caliber 889 with quick date reset, introduced in 1983, demonstrated the company's clear commitment to the mechanical watch. The new *belle horlogerie* era began in 1989, with the Grand Réveil, an exceptional and luxurious alarm watch with perpetual calendar and moon phase. It is powered by 350-part Caliber 919. Its melodiousness comes from a bronze alloy which dates back to the ancient Ming Dynasty Chinese and creates a special sound quality. Floating on the movement, a small clapper sounds the gong, for just under 20 seconds.

The next big bang is the Géographique, a world time watch which only has to be set to a city's name, for the local time and a day/night indicator to immediately appear on the second dial at the six. In 1991, right on time for the 700th anniversary of Switzerland, the Reverso turned 60, and this date was commemorated with a limited edition of 500 watches, the Reverso 60eme with Grande Taille case. This initiated a series of limited edition Reverso modulor in 1000, a tourbillon; in 1994, a minute repeater; in 1996, a chronograph with retrograde display; in 1998, a world time watch; and in 2000, a perpetual calendar. With the Master Control 1000 Hours, Jaeger-LeCoultre presented a watch in 1992 that had to undergo a relentless series of tests before it was allowed on any customer's wrist. A golden seal on the case back visibly certifies this new level of quality. The Master Compressor Diving Pro Geographic brought a diver's watch on the market that featured an integrated depth gauge and a second time zone, and allowed a diving depth of 300 meters. Inside ticks the JLC Caliber 979.

Gyrotourbillon I is another horological masterpiece. With this watch, they came to the idea of taking the ever-changing position of the balance—to compensate for positional errors—to the extreme, because the Gyrotourbillon not only spins around itself in one position, but also rotates on three axes. The watch movement works at 21,600 vph. The Gyrotourbillon returns to its starting position only after two minutes. The movement of the tourbillon cage, together with the escapement, works, as one watch columnist wrote, "like a roller sideways." The inner cage rotates at the same time at two-and-a-half revolutions per minute. If you were to freeze the individual stages of motion, it would create a perfect sphere. Even more than greater precision, this watch offers an incredible spectacle of precision engineering. When you take it off, you can get an even better view of how the Gyrotourbillon works through the case back transparent window. A true Grande Complication, it features, besides the rather distinctive tourbillon, a perpetual calendar with double retrograde date display and a direct equation of time display. Here, the deviation of the mean solar time, which is defined mathematically, is indicated by the actual solar time.

Even for middle-income watch lovers, Jaeger-LeCoultre has one or two tidbits in its watch range. They offer an automatic column wheel chronograph with the JLC Caliber 751, ticking the time at 28,800 vph with ceramic rotor bearings. At a price range several levels higher, but—compared to the competition in their own country—still very reasonable: the Master Tourbillon in steel case, powered by JLC Caliber 978. At this time, barely anyone can offer greater competence in watch movements than Jaeger-LeCoultre.

The Reverso Tourbillon is, like the many other complications in this rectangular watch, a signal of who is undisputably at the leading edge in this métier.

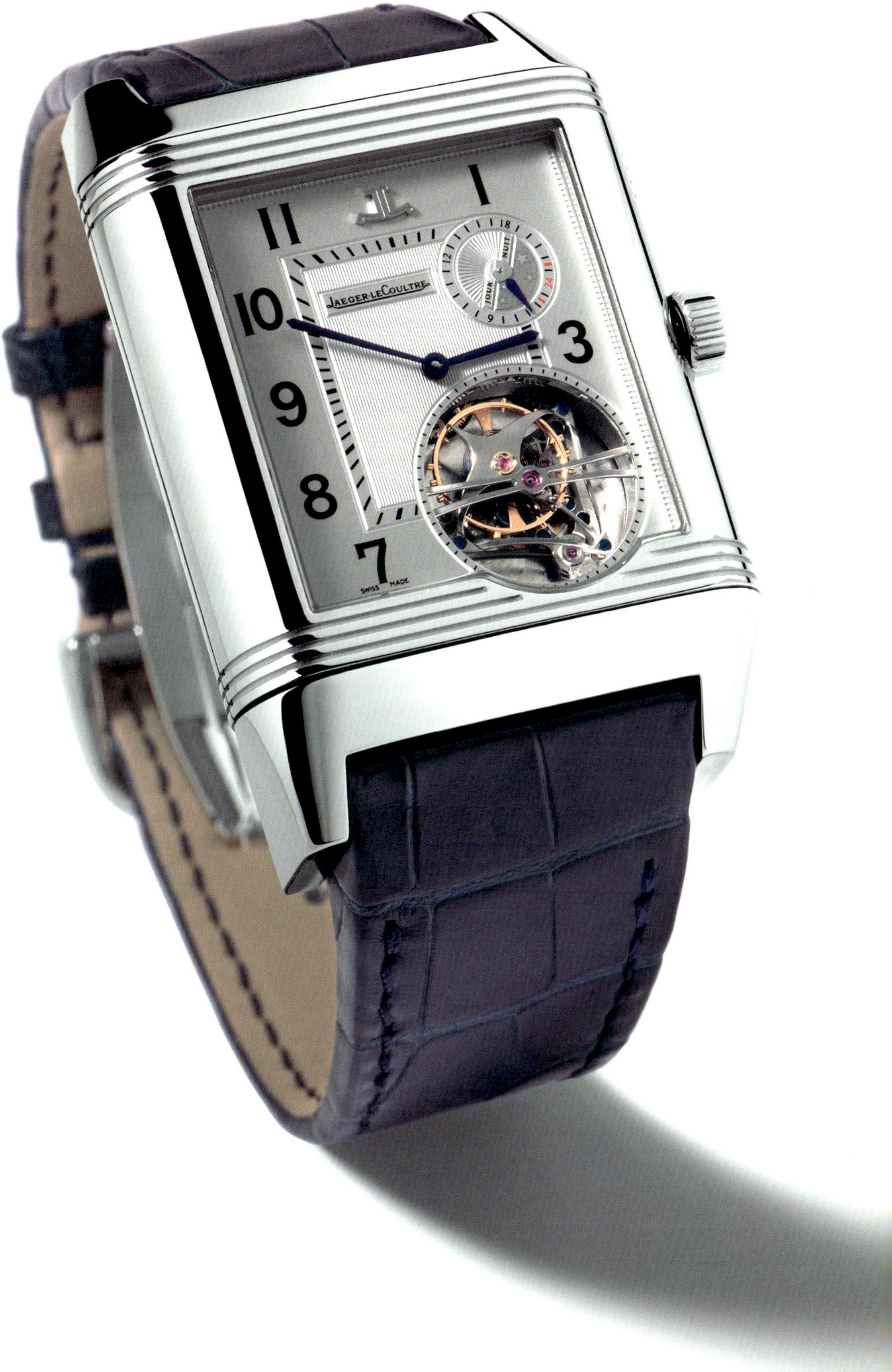
Jaeger-LeCoultre
JOUR
NUIT
SWISS MADE

TACHYMÈTRE
OMEGA
Speedmaster
PROFESSIONAL
FROM THE MOON TO MARS

Complications and Variations

The Pilot's Watch—Not Just for Aircraft Pilots

Long before the world was motorized, people used balloons and airships as the first vehicles to transport them through the air. In 1783, brothers Joseph-Michel and Jacques-Etienne Montgolfier launched a hot air balloon, which rose to 2,000 meters and sailed along for 2 km before it sank back to the ground. But it was only after the invention of the small, fast-running internal combustion engine by Daimler and Maybach in 1883 that the Leipzig bookseller Dr. Wölfert succeed in equipping his dirigible balloon with an effective means of propulsion. The balloon made its first flight from Cannstatt to Kornwestheim in 1888, powered by a Daimler engine. It was Count Zeppelin, with his concept of a rigid airship, who finally perfected transport by lighter-than-air vehicles. Count Zeppelin's first airship, the *LZ 1*, was launched on July 2, 1900, at Lake Constance, powered by two Daimler 10- and 12-hp four-cylinder Model N engines. The era of these gentle giants had begun, and would continue, until the beginning of World War II. It was the Daimler and, eventually, the Maybach engines that gave these giants of the sky the power to travel over oceans and continents. It was not until 1903 that aircraft finally took off into the air, if only for short distances. The German Wilhelm Kress already had built a seaplane in 1901 with a Mercedes engine, but unfortunately he was not able to take off in it. That was not achieved until two years later by Karl Jahto, but the flight was uncontrolled. The American brothers Wilbur and Orville Wright were able to succeed in making the first steered and controlled flight. On Dec. 17, 1903, Orville Wright flew for twelve seconds in the 2-hp four-cylinder engine equipped *Kitty Hawk*, covering a

After air flight came space flight. Omega and the Speedmaster, which can truly be grandly advertised as: "The only watch worn on the moon." A robust manual winding chronograph, which still wears well today.

distance of 37 meters. His brother Wilbur was able to fly for almost a minute over 260 meters. This launched airplanes into a competitive battle with airships, and to ultimately assert themselves in the battle for the skies.

With the advent of air travel, the watch also became an ever more important tool for aeronauts and aviators. Here again, the wristwatch, with its functionality, showed its superiority to the pocket watch—which perhaps could be easily used by a commander standing on the bridge of a zeppelin, but not by a pilot hindered by his position sitting in the cramped cabin of an aircraft.

One watch company that became extensively engaged in manufacturing special pilot's watches, was the International Watch Company. IWC Schaffhausen introduced their first special pilot's watch in 1936. To operate

IWC's 50000 Caliber Big Pilot's Watch with Pellaton automatic winding, seven-day power reserve, and date display at the six (above). The Lindbergh Hour Angle watch by Longines. During the early years of aviation, pilots would navigate using the bi-directional rotating bezel with hour angle increments (right).

successfully in the cockpits of contemporary aircraft, special features were needed: an optimally readable black dial with prominent luminous hands, large luminous numerals, and a rotating bezel with register hand for short-term measurements. These principles still apply today for how these watches should function. Based on these special pilot's watches, IWC made the Big Pilot's Watch in 1940 to military specifications, with original

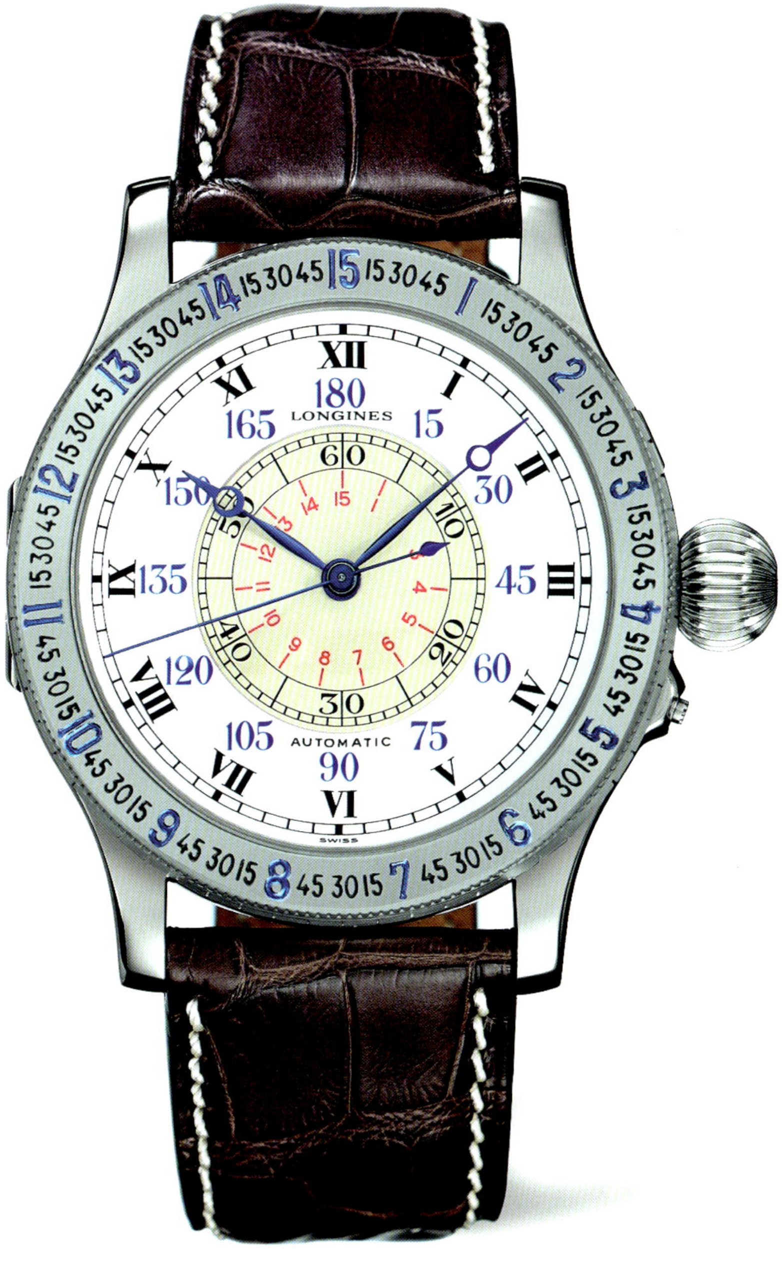
LONGINES
AUTOMATIC
SWISS

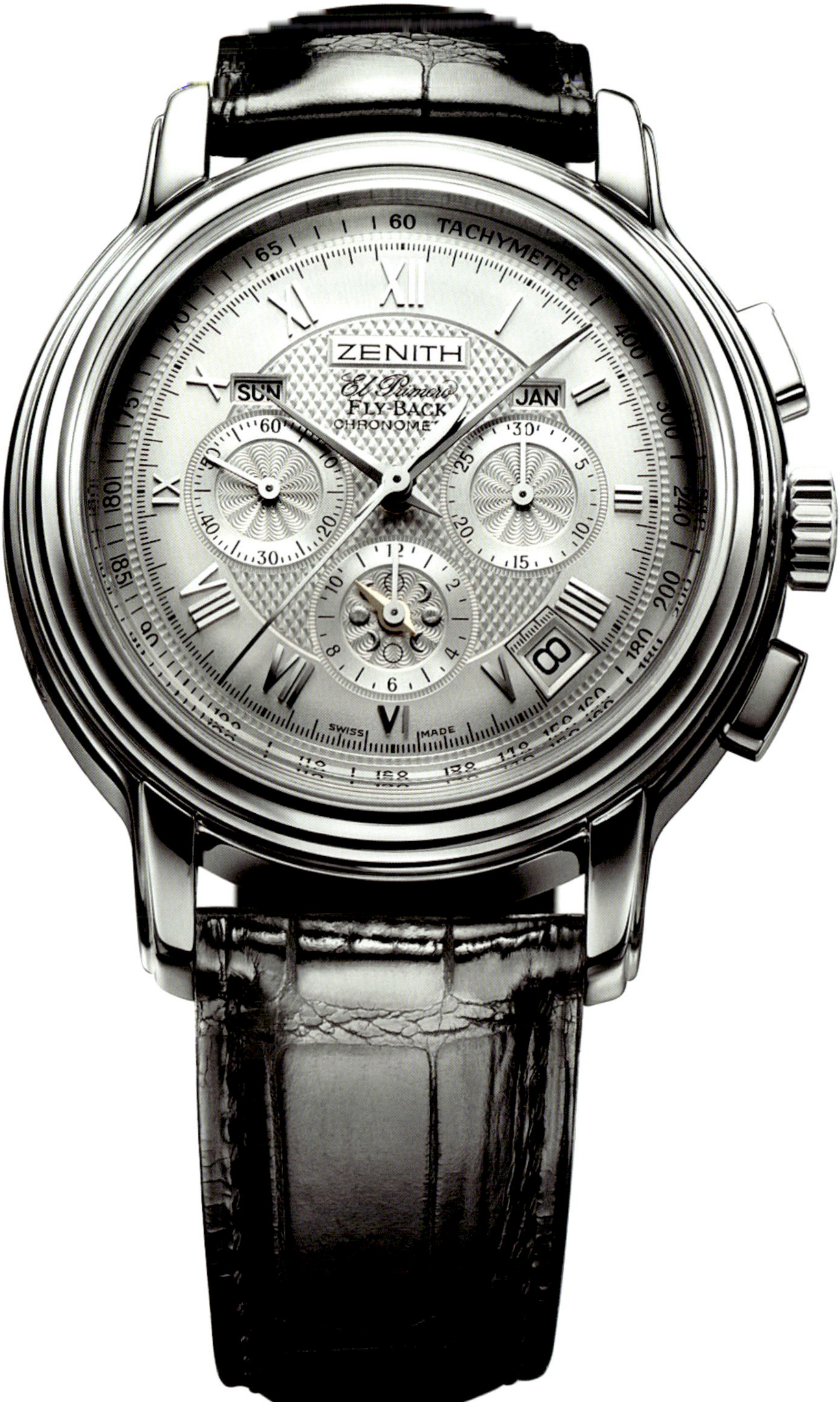
ZENITH
El Primero
FLY-BACK
TACHYMETRE
SUN
JAN
SWISS
MADE

pocket watch movement and large central second—a certified observation and navigation watch for military pilots.

The most famous IWC pilot's watch, the Mark XI with manual-winding Caliber 89, was used for both civilian and military aviation from 1948. Its advantage over other pilot watches: it had an additional soft-iron inner cage for protection against magnetic fields. The new Pilot's Watch Classic collection follows the tradition of these timekeepers specially designed for pilots. It comprises five models: the new Big Pilot, the Double Chronograph, the Chrono-Automatic, the classic Mark XVI, and, as an innovation in the pilot's watch range, the Midsize model with a diameter of 34 millimeters.

Both Junghans, with its legendary Caliber 88, and Heuer, with the Valjoux 230 chronograph, make watches for the pilots of the German *Bundeswehr*.

Inspired by these classic pilot's chronograph models, the Frankfurt-based Sinn company also manufactures this type of watch as a rugged everyday watch. One of the classic models is now being manufactured once again, due to popular demand. The watch enhances the 103 model range with its black dial with white totalizer rims. Sinn was already making this model about ten years ago, with a manual-winding movement. The 103 A Sa model, however, is equipped with an automatic Valjoux 7750 movement, and has a stainless steel case and stainless steel pilot's ring with black anodized aluminum inlay. Trendsetters in the field of pilot's watches include Breitling and Tutima, as well as Fortis, which usually offers chronographs with or without rotating bezel that allow the sports flyer to make short-term measurements.

Actually more a calendar watch, but with an essential feature often found in pilot's watches, this Zenith fast beat watch has a flyback function (left). The Breguet also has a flyback. This red-gold chronograph has a central minute counter hand (top right).

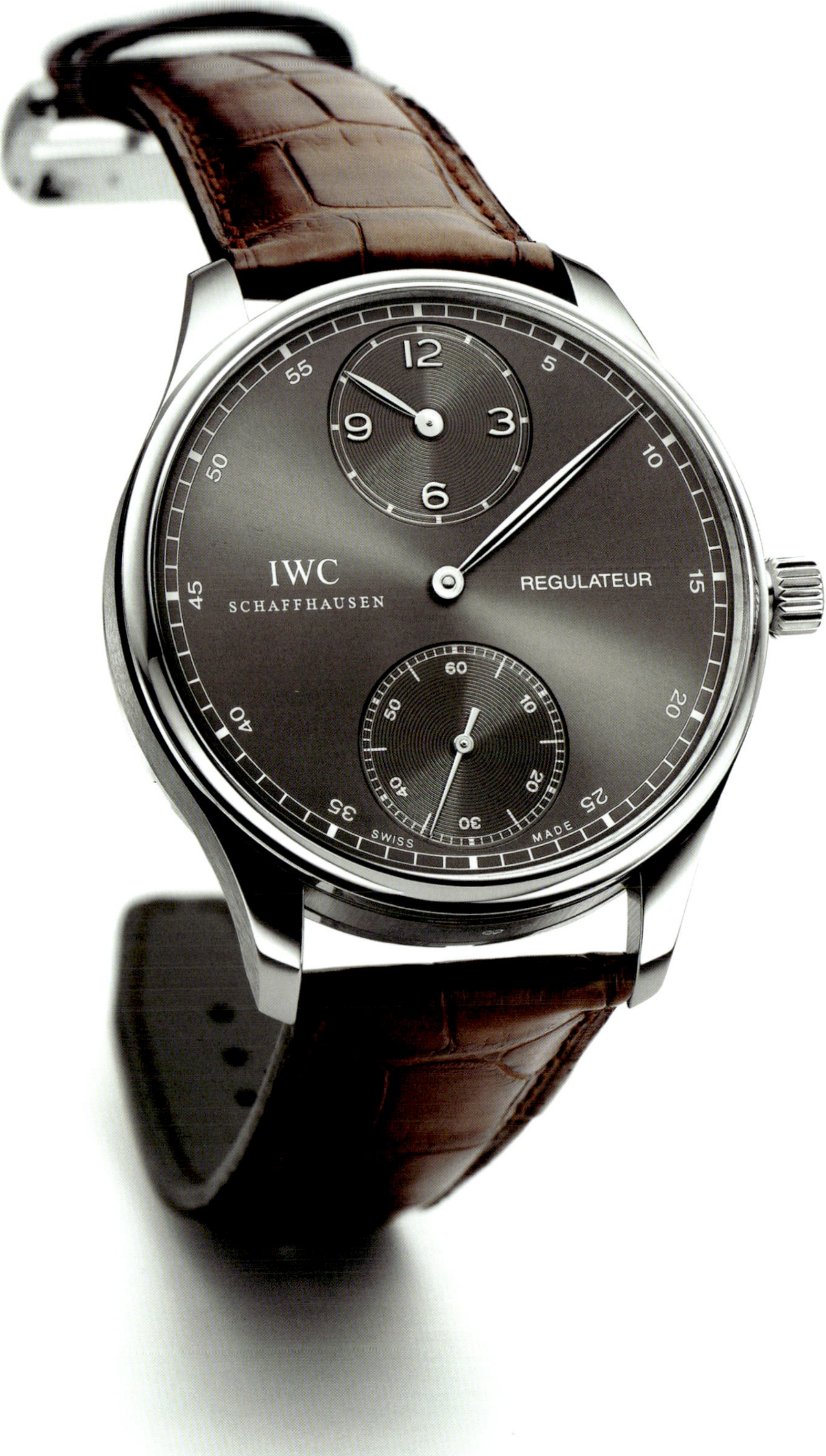
IWC
SCHAFFHAUSEN
REGULATEUR
SWISS
MADE

The Regulator—Precision is the Highest Precept

This exceptional dial design, in which all three time displays are separate, was invented at the end of the 18th century by French watchmaker Louis Berthoud. He presented a marine chronometer with decentralized hour hand, which prevented the hour hand from covering over the second or minute hands, ensuring optimal readability. Precision clocks, as used by observatories, the post office, railway, and watch manufacturers preferably featured a regulator dial, for the sake of exact readability. This is true of the 1927 precision pendulum clock by Clemens Riefler (Munich), which also had a 24-hour display, or the clocks by Strasser & Rohde and Knoblich. Sattler of Munich continues to offer such clocks today, which have an accuracy of two seconds per month and thus are more accurate than a conventional quartz watch.

In 1987, Gerd-Rüdiger Lang introduced the regulator dial for the wristwatch—a limited edition with a manual-winding Unitas movement. Although the Italian Leonardo Spinelli had already made two wristwatches with regulator dial in 1960, this wasn't used as a basis for serial production. As a result, Chronoswiss was the first watch brand to offer this feature to the broader public. In the meantime, Chronoswiss has been offering this unusual dial design along with different complications such as tourbillon and chronograph.

To present this unusual single push-piece chronograph with column wheel control, Lang also created a new name: Chronoscope, which comes closer to the meaning of its

IWC also offers the Portuguese with regulator dial. Inside ticks a manual winding movement with a long fine adjustment push-piece and 3/4 train wheel bridge (left).

Also manual winding and a comfortable 18,000 vph: a Régulateur in rose gold from the Frankfurt-based Sinn company, direct marketer of these high quality and yet affordable watches (right).

VACHERON CONSTANTIN
GENÈVE
TOURBILLON REGULATEUR
SWISS
MADE
VACHERON CONSTANTIN
P30080
MADE

function, to see the time, than the German term *"Zeitschreiber"* ("chronograph"), which literally translates as time recorder. Besides Chronoswiss, other companies offer a tourbillon with regulator dial, such as Ingersoll's Charleston model, very affordable at $13,500, and Chopard—of course, considerably more expensive. Today, regulator dials are also available on diving watches, such as the Oris Big Crown Divers Regulator. This watch, waterproof to 200 meters, is powered by an ETA 2824-2 and has all the features necessary for diving watches, such as a uni-directional rotating bezel and screw-down crown. Even such futuristic watches as the Mercedes SLR McLaren Chronograph come with regulator displays. Some manufacturers such as Guinand offer their Régulateur with central pointer date display, which is, however, more difficult to read. Better are the complications such as the moon phase, available from Goldpfeil Genève, or the Chronoswiss Tourbillon Squelette.

Rüdiger Lang, father of Régulateur dial for wristwatches, also makes this skeleton watch with tourbillon and Régulateur dial that allows an unobstructed view of the movement workings.

A Tourbillon with regulator dial from the oldest watch company in the world. The manually wound movement has a power reserve display at the eleven.

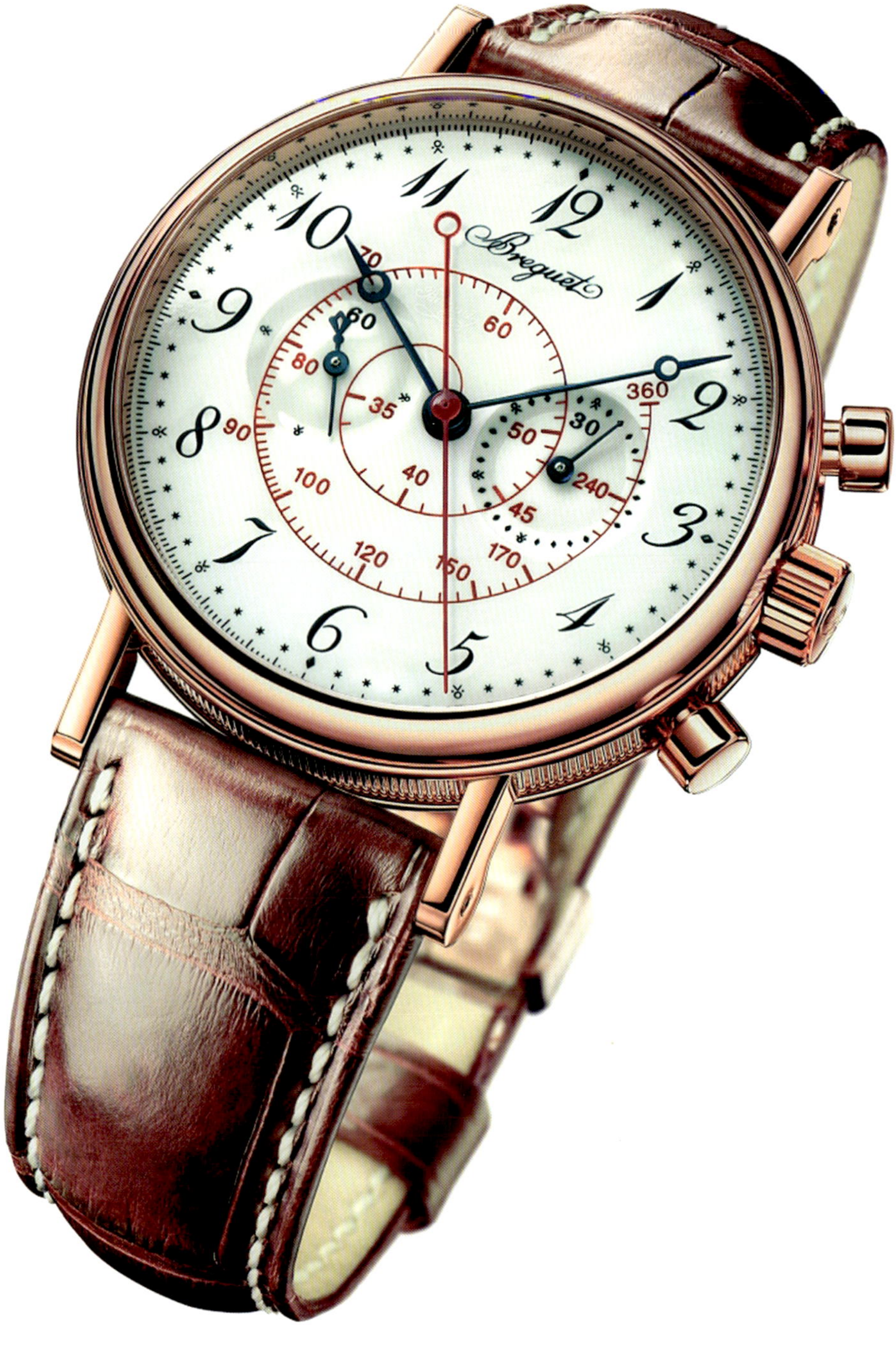
Breguet

The Chronograph—The Sportsman among Watches

It was the complications which, above all, helped create a breakthrough for the renaissance of higher-price-range mechanical watches. Here the chronograph—not to be confused with the chronometer—was most important. The "time recorder," the literal translation of chronograph, was in fact first devised by French watchmaker Rieussec in 1820 in Paris. A tiny drop of ink was applied to a watch dial using a pen, to mark the beginning of the stop-watch time measurement.

In 1831, a former employee of Abraham-Louis Breguet, the Austrian Joseph Thaddäus Winnerl, presented a watch with *seconde indépendante*, i.e., independent seconds. This design allowed the operator to stop the second hand arbitrarily and start it again without affecting the watch movement itself. The down side was resetting the seconds to zero, which took up to a minute after setting the hand. The basic invention which made zero-reset possible was the reset hammer, a mechanism resting on a heart-shaped cam on the tube of the stop-seconds hand. To reset the chronograph, a spring-loaded lever jumps it against the curved cam and immediately resets the hand to zero. The hammer was invented in 1862 by Adolphe Nicole, who was working in the Vallée de Joux in the French region of Switzerland. That was the birth of the modern chronograph. First as single push-piece chronographs in a pocket watch, then later as addition timers, miniaturization eventually made it possible to adapt the chronograph for wristwatches.

For any fan wearing a chronograph on their wrist, its inner workings play a significant role. The connoisseur would prefer a column wheel control for the chrono function, and there are currently only a few high quality alternatives for a modern automatic movement. They include the Zenith El Primero Caliber 400, the first automatic chronograph movement in the world, the new Rolex chronograph movement, an automatic chronograph from Patek Philippe, and finally, the Piguet 1185. This is one of the movements sold only in small numbers to renowned manufacturers such as Breguet and Blancpain. Because of its modular design—it can be manufactured as a manual-winding, automatic movement, rattrapante, or flyback—it is often mistakenly advertised as a modular caliber. This is of course wrong. It is a highly

A really classic chronograph from Breguet, an addition timer with snail-shape tachymeter scale.

exclusive thoroughbred automatic chronograph movement. Due to the design, with a diameter of 25.6 mm (11½ lignes) and a height of 5.4 mm, it is also the smallest and flattest movement of this type. The movement has 37 jewels and a vibration frequency of 21,600 Hz. Using a micrometer screw, it can be finely adjusted to five positions and for different temperatures. Overall, this little marvel of technology is made of 308 individual parts. Some movement parts are decorated with colimaçon, others, like the solid gold rotor, with Geneva stripes. Blancpain features a 21-carat gold rotor to wind the watch. The first chronograph movement with automatic winding and column wheel control was made by Zenith in 1969 and is still state-of-the-art. Rolex, Patek Philippe, Seiko,

The IWC Da Vinci with its own automatic chronograph movement and column wheel control. Other features include the date at the six and flyback function. Here in platinum, rose gold, and steel.

Classic beauty: Reference 5070. Platinum manual winding chronograph with a control wheel caliber made by Nouvelle Lemania exclusively for Patek Philippe.

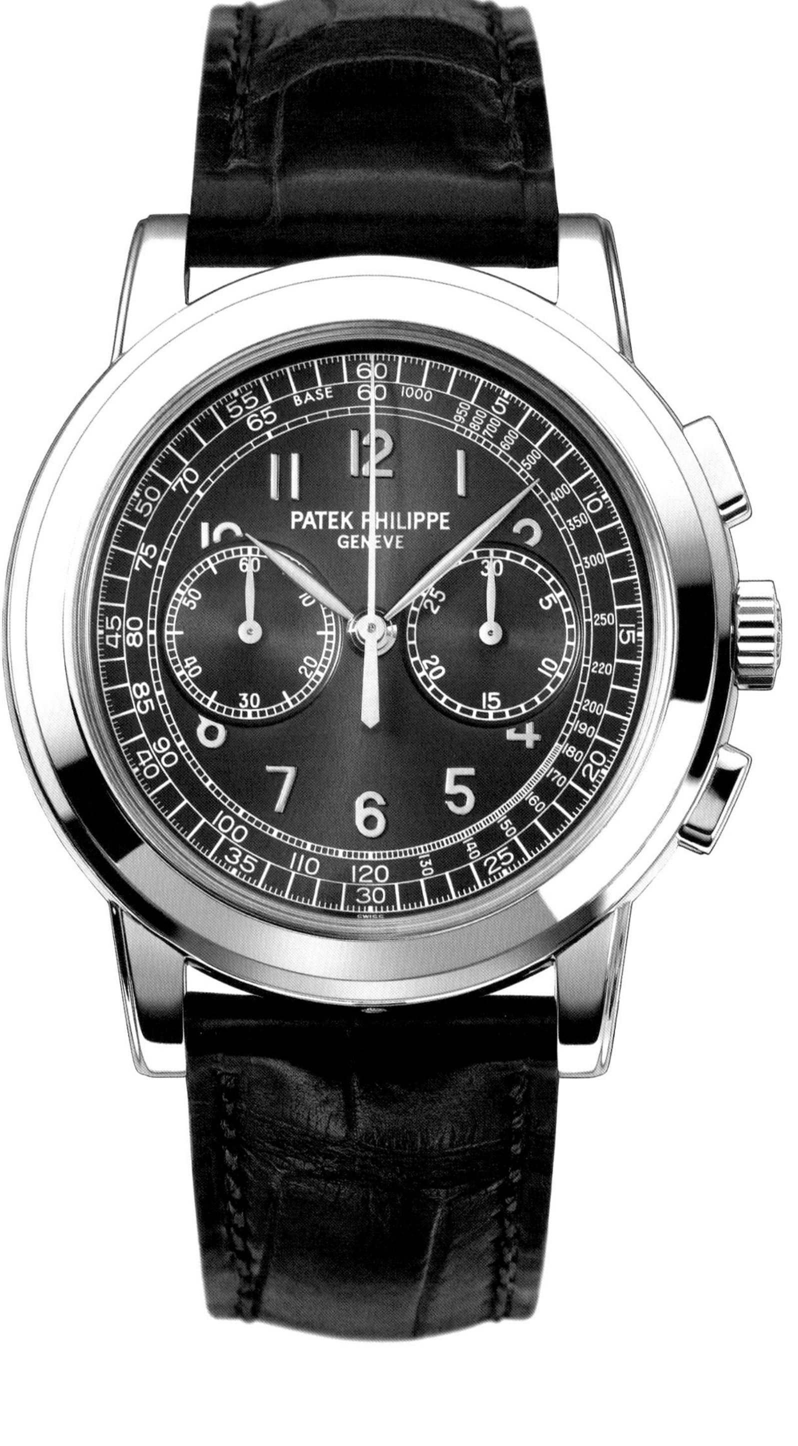
PATEK PHILIPPE
GENEVE

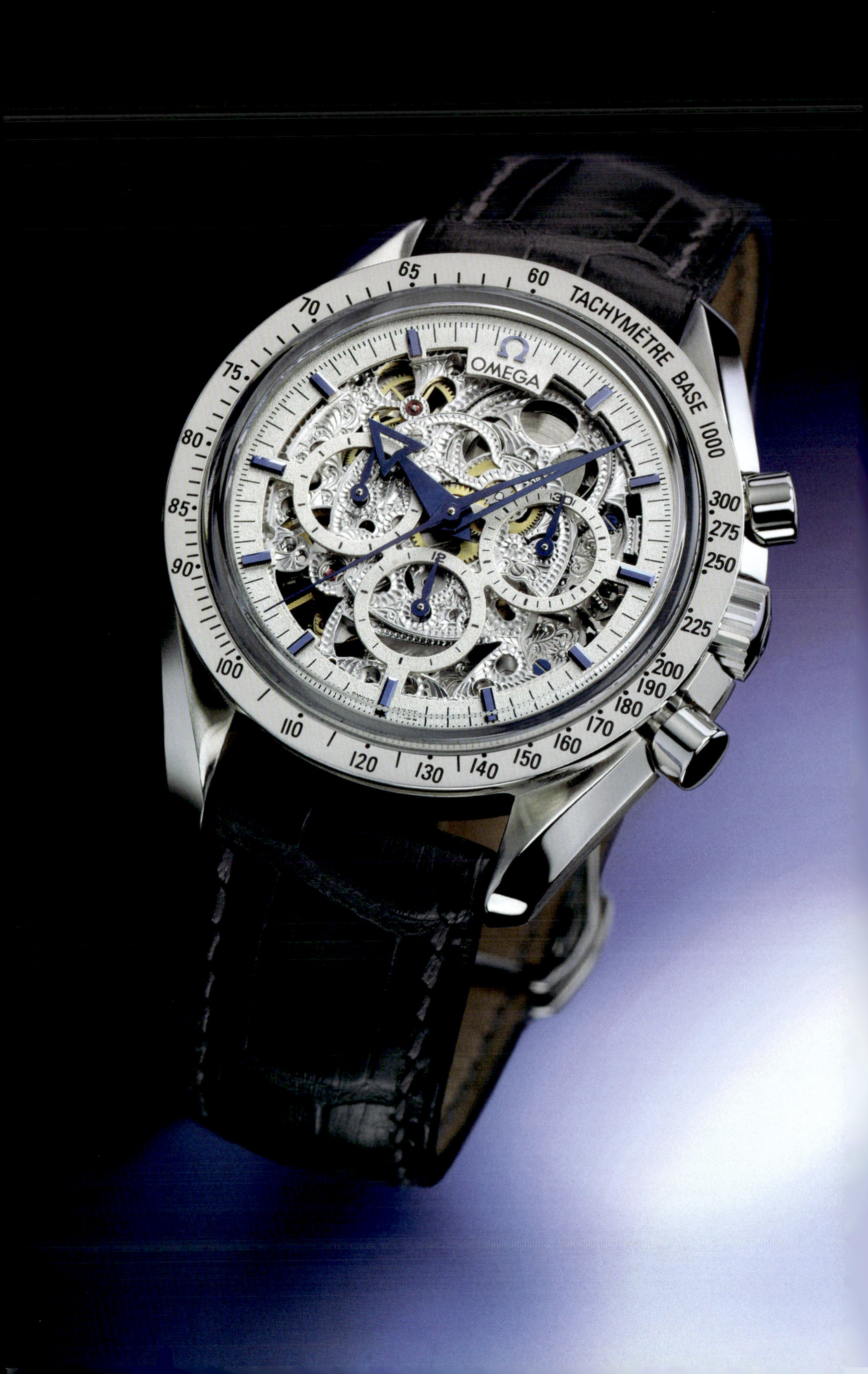
OMEGA
TACHYMÈTRE BASE 1000

Blancpain, and Omega all also offer automatic column wheel control chronographs. The vast majority of chronographs available today are addition timers. The push-piece at the two activates the chronograph and also stops it; the push-piece at the five resets the stopwatch hand and totalizers to zero. There are also a range of special features, such as the flyback function. This construction principle was primarily used for military aviation chronographs. For the German *Bundeswehr*, it was the Valjoux 230 which powered the Heuer or Leonidas chronographs. The special feature of this quick reset, is that stop watch timing can be re-activated immediately without interruption, by pressing the bottom push-piece. This eliminates the time-consuming stop and reset of the chronograph hand. This function, also called *retour en vol*, was used to create the Caliber F185 from Caliber 1185. At present, only two other manufacturers offer this complication. These are Zenith, also with a column wheel chronograph, and Breguet. The latter maker uses a cam/lever switching control. Another classic complication is the rattrapante or split-seconds chronograph. Rattrapante, derived from the French verb *rattraper*, means something like recapture, recover, recoup, catch up on, and describes this type of control better than the German concept of the *Schleppzeiger* (the English term is split-seconds; a more literal translation would be drag indicator or trailing pointer). Because the pointer or hand is "dragged along" and possibly uncoupled or released, at any time it is possible to again catch up with or overtake the second hand, something not possible for a split-seconds.

Often upgraded with this complication by different manufacturers, the Valjoux 7750 is currently most used automatic chronograph movement, along with the stripped down for manual-winding 7760 version. While many manufacturers set a single pliers mechanism out of sight under the dial in rattrapante chronographs, IWC's Portuguese Rattrapante model offers a beautiful double pliers design on the back of the movement. It thus resembles the structure of the Venus 179, which was made by Breitling, among others, in the 1950s, even though the control of the chronograph is by cam/lever design and not column wheel. This gem—column wheel, automatic and rattrapante—is today still done by Blancpain.

No other watch design fascinates so many male watch aficionados as the chronograph. It is the only complication that allows the wearer to intervene in the watch movement themselves. Of course, you can set an alarm or call up the time audibly by the minute repeater, but both are one-time actions. A chronograph, in contrast, allows the user to play with time like a virtuoso, measuring it divided into seconds, minutes, or hours, or interrupt the measurement—you can be in a constant dialogue with the

The Omega Speedmaster as an elegant skeleton watch in white gold. You can also wear a chronograph this way.

chronoscope (a more exact term for the chronograph). Chronoswiss chief Rüdiger Lang was one of the first to be terminologically correct and use the term chronoscope, and this for an unusual watch: a single push-piece chronograph with regulator dial, controlled by a column wheel below the dial. To achieve this, the chronograph mechanism was constructed in an Enicar Caliber 165 movement, an in-house caliber that is the basis for the Régulateur Chronoswiss as the C 122.

Another special feature in the chronograph landscape—which is hardly lacking in special features—is the *foudroyante* or *diablotine*—flying seconds in English—a mechanism that allows reading the fractions of a second on a sub-dial. Depending on the balance frequency, it will rotate five, six or even ten times a second. A German term for this function is *blitzenden Sekunde* (lightning seconds)—in English, jumping seconds is also used. For this complication also—only offered by a few manufacturers such as Girard-Perregaux in the Vintage Foudroyante XXL or the Graham Foudroyante—the proven Valjoux 7750 provides the impetus. Its 28,800 vph makes it possible to measure to the eighth of a second. In most cases, it is linked to a further complication—in these two pieces, a rattrapante.

A normal chronograph with central seconds and 30-minute display by a small totalizator already offers, depending on its scaling, any number of measurable events: the telemeter scale, to show how far away a thunderstorm is, the tachymeter to determine your speed by foot, bicycle, or car, the pulse monitor and unit counter, up to counting telephone call charges. But even without such dedicated scales, a chronograph offers useful additional services besides measuring time. The central indicator atop the minute or hour hand, offers the viewer an easily comprehended overview of elapsed time. An absolutely essential appointment can be tagged by setting the stopwatch hand on the appropriate time. If you have a rattrapante, you can even set two events for your attention in one day. No other complication adapts so well to linking with other complications as the chronograph. The calendar, perpetual calendar, alarm, tourbillion, and world time indicator, all combine readily with a chronograph.

Patek Philippe's first in-house chronograph caliber with column wheel control. It also features an annual calendar and power reserve indicator. This watch is available in platinum and rose gold.

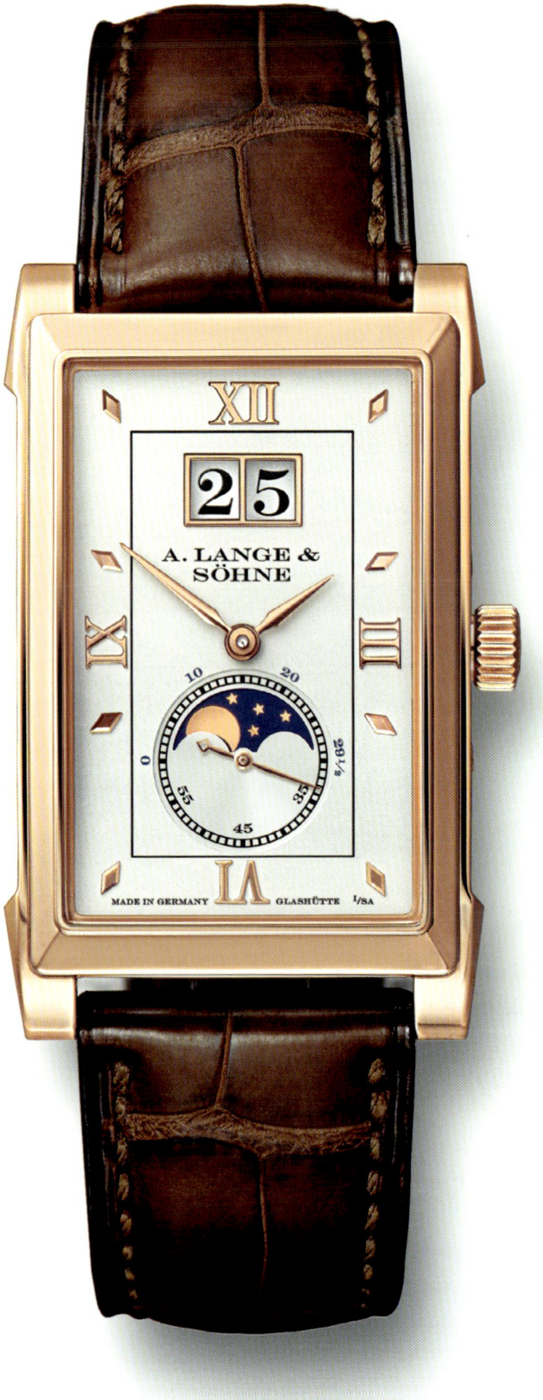
25
A. LANGE & SÖHNE
MADE IN GERMANY
GLASHÜTTE

The Moon Phase—The Moon's Disc on your Wrist

The mechanical watch renaissance owes its origin not least to the fascination exerted by a moon phase display—that little moon which adorns the dial as a smiling face or a golden disc. Blancpain presented such a model in 1984, which immediately won customers' approval. For all those for whom the moon plays a special role in their lives, even when skies are cloudy, they can ascertain the particular phase of the earth's moon anyway.

At a distance of 363,000 to 406,000 kilometers with a diameter of 3,476 km, the earth's small satellite moves in its orbit around the blue planet. The exact time it takes the moon to orbit is 27 days, seven hours, 43 minutes, and 12 seconds. The so-called lunation, or lunar phase cycle, which is also affected by the earth's motion around the sun, is exactly 29 days, 12 hours, 44 minutes and 2.8 seconds. During this time, the moon passes through its phases of new moon, first quarter, full moon, last quarter and new moon again. Watchmakers have made things a little easier by setting the orbit exactly to 29 days and twelve hours. Here we find the reason for the two moons of the lunar disc, which is switched daily. It is easier to represent 59 days than half that number. After two and a half years, it must be manually corrected by one day.

However, there are designs, such as from A. Lange & Söhne, where the moon phase keeps moving along continuously, reducing the display error to 1.9 seconds. Only the second generation inheriting this watch, given an average life expectancy, will have to correct the moon phase by hand, in 122 years.

The moon phase usually is available in combination with other complications, such as the full or perpetual calendar. The

In a shaped movement specially created for this manual winding watch, A. Lange & Söhne presented Big Date and moon phase on the Cabaret (left).
This Lange movement readily shows us that the moon disc mounted on the movement's upper side includes two moons (right).

Pocket watch movement with moon phase. The IWC Reference 5251, the inspiration for the Portofino line, was reissued for the Vintage collection and is again available in steel and platinum as Reference 5488 (above).

A Blancpain calendar watch with moon phase. The Caliber 6763 is automatic, and can be admired through a sapphire crystal display back (right).

conceptual design used for most moon phase watches is the disc described above, but there are other variations, such as a rotating moon that moves around the dial, as on the Moon Time I by Bunz, or the representation of the different moon phases swept over by a hand, as on one Patek Philippe model. However, most moon phase displays are correct only for the northern hemisphere. Only the IWC Portuguese Perpetual Calendar shows both lunar phases, for the northern and southern hemispheres, displayed in two fields. A. Lange & Söhne also makes this special feature. The consumer has to purchase a double set of two watches. The Luna Mundi as the Southern Cross, in rose gold with brown crocodile leather strap, and as the Ursa Major, in white gold with black crocodile leather strap, each correctly display the moon phases for both hemispheres. The moon phase discs also each display the constellations Ursa Major—the Big Dipper—or the Southern Cross. For the southern hemisphere display it was only necessary to insert an intermediate wheel which reverses the direction the moon phase display rotates.

As with calendar or perpetual calendar watches, on a moon phase watch it is difficult and unnerving to reset the watch after it has been allowed stop. Using a watch winder will keep an automatic watch running. For manual watches, the Orbita company offers a device that winds it daily by the winding crown. The device has a microprocessor to continuously measure the tension of the mainspring and control the winding process.

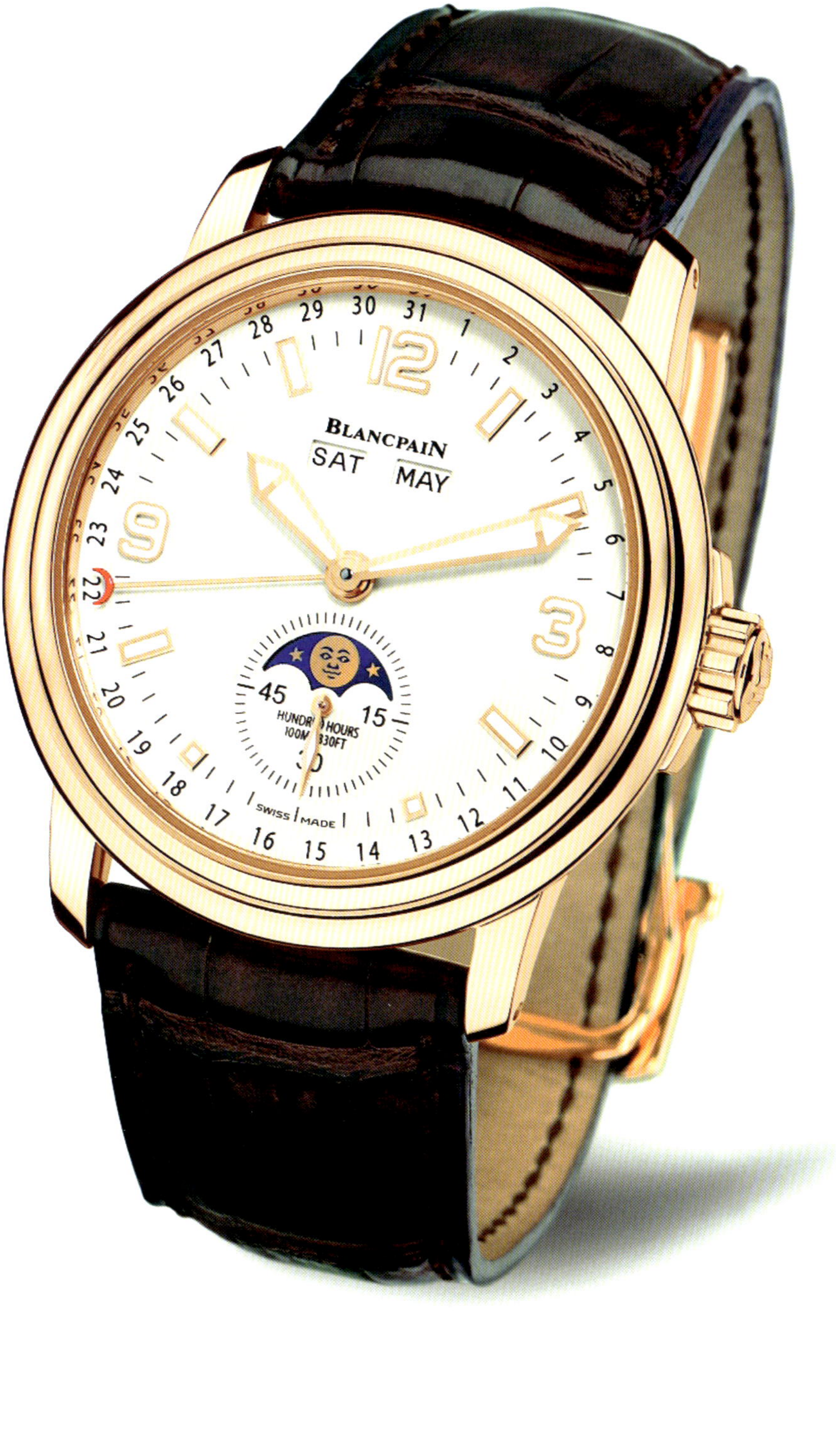
BLANCPAIN
SAT
MAY
SWISS MADE

In any case, such a device costs many times what an ordinary watch winder costs.

Already early on in the history of the mechanical clock, the moon phase was a popular complication, and more and more people delighted in the changing moon disc on the dial. The Belgian Henry already had made a pocket watch with moon phase and calendar display by 1650, which shows that people have been appreciating this complication for over 350 years.

Moon phase and big date with a difference. The Glashütte Original Panorama Date with its asymmetric dial (above).

The Valjoux 7751 is, like its manual predecessor, the Valjoux 88, a chronograph movement with calendar and moon phase. A robust movement, which also allows including these complications in a sports watch (right).

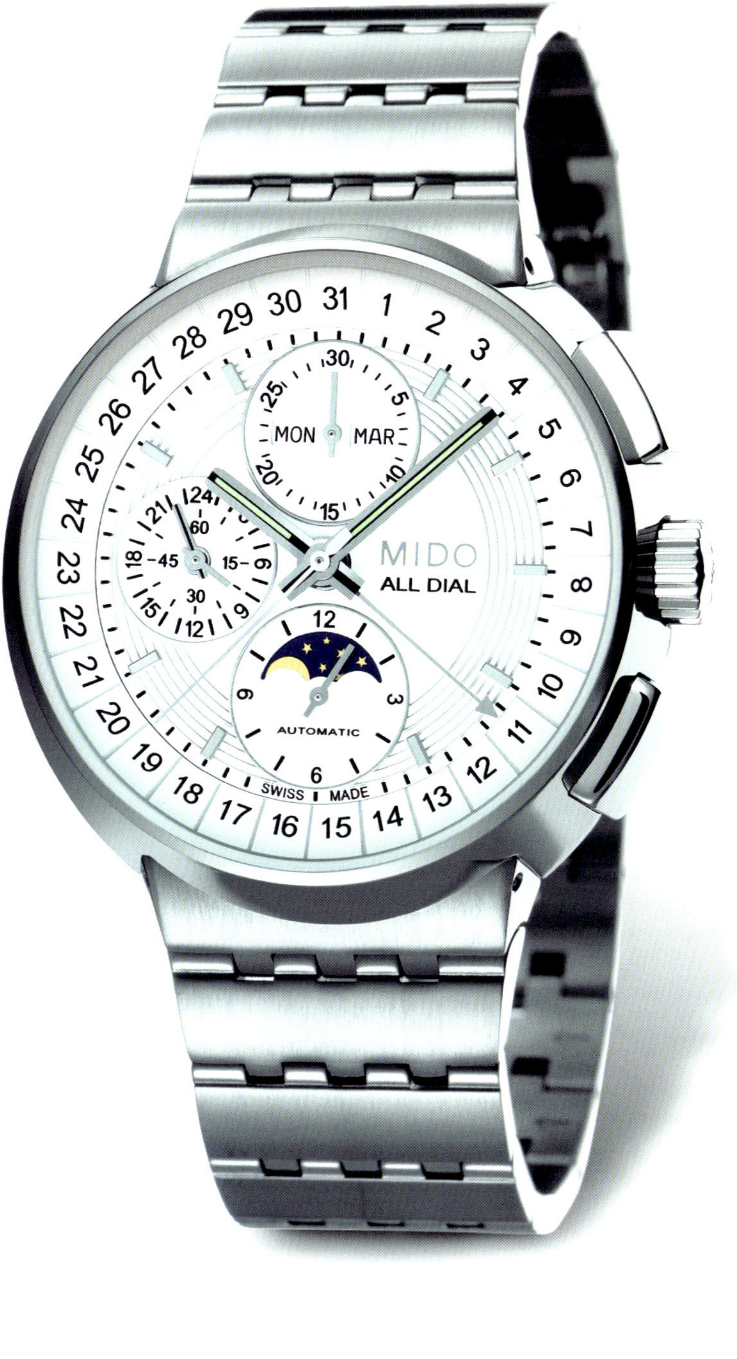
MIDO
ALL DIAL
MON
MAR
AUTOMATIC
SWISS
MADE

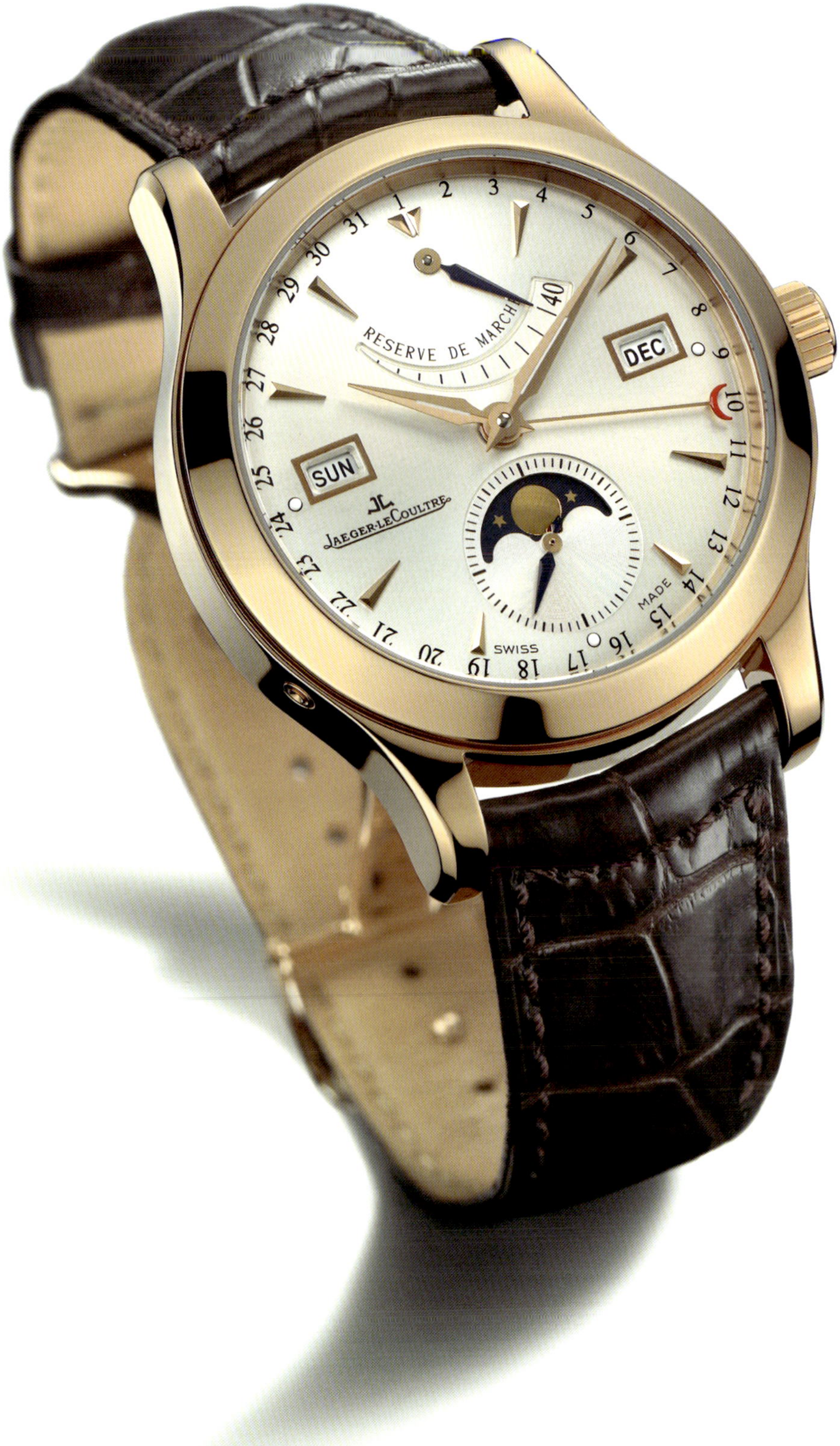
RESERVE DE MARCHE
40
DEC
SUN
JAEGER-LECOULTRE
SWISS
MADE

The Calendar—Day and Month on the Dial

To understand the specific problems a calendar display creates for a watchmaker, it is necessary to take a brief look at the development of our calendar. Originally, and still today, the basis of our calendric recording of time is the Julian calendar, which goes back to Gaius Julius Caesar, and replaced the ancient Roman calendar, based on the moon. This had a year of twelve months with a mean length of 29.5 days. It always required corrections, because the solar year ran about ten days ahead of the calendar. Based on the calculations of Hipparchus of Nicaea, who had already described quite accurately a solar year of 365 days, 5 hours, 55 minutes and 12 seconds in the second century BC, a new calendar was created which began on March 1, with months alternating between 30 and 31 days. Only the final month of the year, February, was a leap month of 29 or 30 days.

Despite initial difficulties—it was necessary to add a leap year every three years, and some modifications initiated by Augustus—to whom we owe August and a leap month reduced to 28 or 29 days—this Julian calendar functioned until the year 1582. Since the Christian Easter festival was moving ever-closer to mid-year, a validation showed that the Julian calendar was exactly 11 minutes and 14 seconds too long, and thus, after 128 years, made a correction of one day necessary. The Gregorian calendar, which goes back to Pope Gregory XIII, came into use and solved the problem of incorrect year length, requiring that every four years a leap day is added to the calendar, except for century years; only those divisible by 400 are to be leap years.

The calendar timepieces, already in use since the 14th century, included a calendar movement linked to the clock movement; these took no account of such subtleties and had to be reset by hand no later than every three months. In simple calendar watches, this has not changed up to the present. The calendar display is either in a window or shown by a hand, either mounted in a small

Jaeger-LeCoultre—proof, that the Master is master of all complications. The Master Calendar in red gold Reference 151242D with JLC 924 automatic movement also has a power reserve indicator.

sub-dial or a central month and day hand. Some manufacturers made a virtue out of necessity, leaving the month display decoupled from the movement, so that it had to be manually reset for each new month.

Despite the widespread use of day/date displays today, a calendar watch is still considered one that displays day, weekday, and month. In the past, they primarily used Valjoux movements, used by many manufacturers as the 72C and 88 as chronograph movements with calendar indicators. The former was a pure calendar movement, while the Valjoux 88 also featured a moon phase. Today, a Valjoux movement, the 7751, with calendar, 24-hour display, and chronograph, is the most widely used; however, as an automatic. This simple but robust Valjoux design has been surpassed by the Zenith El Primero movement; as the 410, it has a full calendar with moon phase. Unlike the Valjoux, it displays the date in a window and not by pointer. Technically, this Zenith Caliber 410, with its column wheel control and integrated calendar mechanism, is one of the finest available as a calendar watch. However, there are also simple movements, such as the ETA 2892, available in their various derivative forms with calendar module—not a bad choice, because this movement is extremely sturdy. Blancpain offers a calendar watch with 100 hours power reserve in the Caliber 6763, and Jaeger-LeCoultre's 89-448-2 is an attractive calendar caliber, which is often used by other manufacturers.

However, all these calendars have the disadvantage described above: they always count 31 days and therefore have to be hand-corrected for the shorter months. In addition to the perpetual calendar, some manufacturers, such as Patek Philippe, offer annual calendars which only need to be corrected in February. The Patek Philippe Caliber 315 S is available both as just an annual calendar watch or with additional moon phase display.

When wearing a calendar watch, it is important to remember that it depends on a rather sensitive technology, not suitable for sport activities. Also, the buyer of a calendar should remember that once they have stopped, it is difficult to set them running again. This makes an automatic calendar watch preferable to a manual-winding one.

A calendar watch by Vacheron Constantin with retrograde day of the week and date. A platinum watch whose value only a connoisseur would recognize.

1 · 3 · 5 · 7 · 9 · 11 · 13 · 15 · 17 · 19 · 21 · 23 · 25 · 27 · 29 · 31
VACHERON CONSTANTIN
GENÈVE
MONDAY
TUESDAY
WEDNESDAY
THU
FRIDAY
SATURDAY
SUNDAY
PT 950
SWISS
CONSTANTIN
86020
VACHERON ✶ CONSTANTIN
SWISS MADE
N°1
1149520

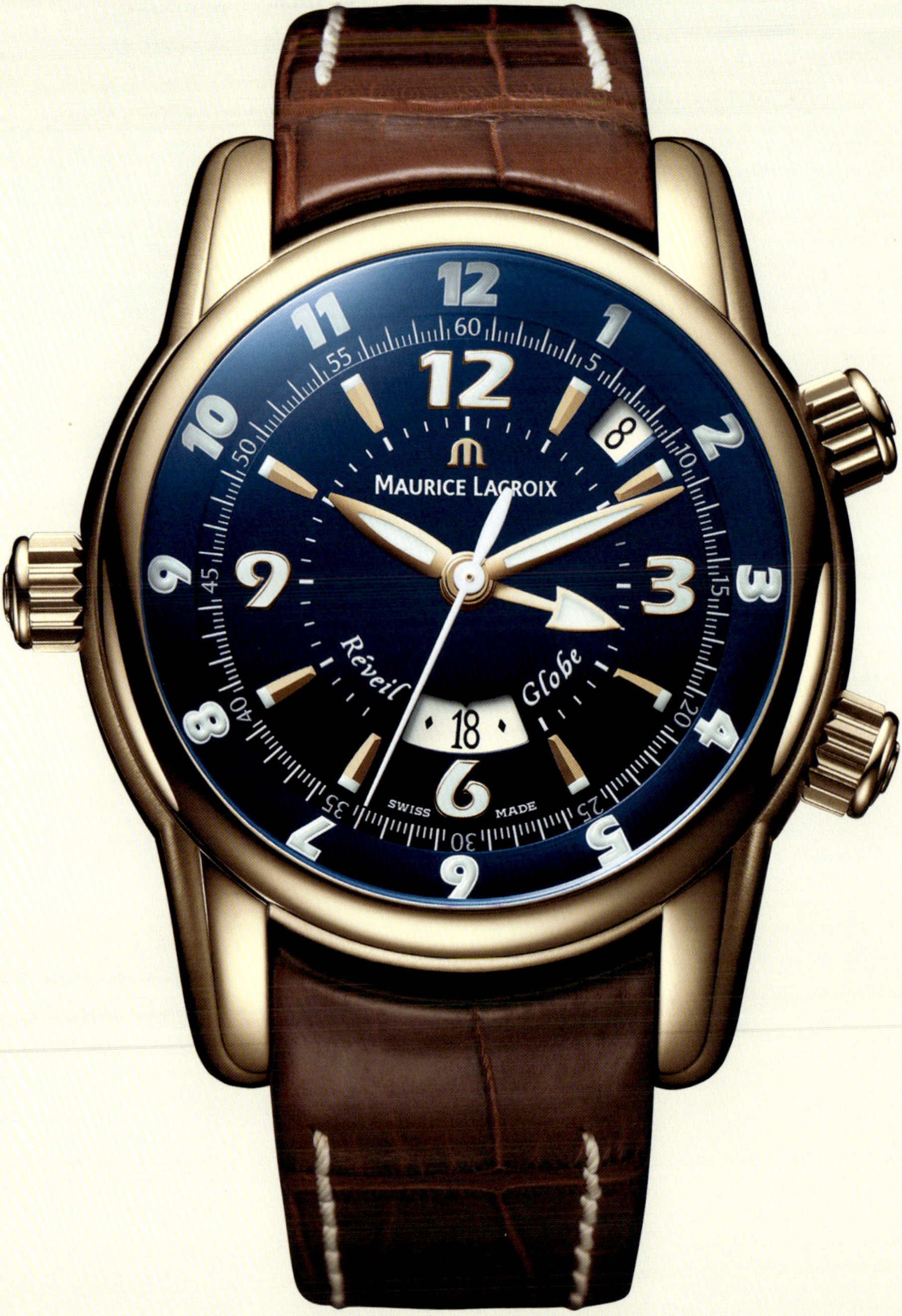
MAURICE LACROIX
Réveil
Globe
SWISS
MADE

The Alarm—Even Small Watches Can Ring Out Loud

Anyone who is tired of the annoying and sometimes embarrassing peeping of electronic technology during conferences may well pine for the good old ringing or rattling of an alarm clock. This might then lead you to get a wristwatch with an alarm, and surprising all your colleagues equipped with exuberant electronics, with the delicate sounds of a mechanical alarm. The alarm: a complication rather neglected until recently and, with one exception, more generally available in less expensive mechanical watches. In the past, an alarm was a component of a table clock, then the pocket watch. Indeed, Eterna presented a 13-ligne alarm movement in 1914, for which they had already applied for a patent in 1908, but there was no widespread acceptance for this innovation. Vulcain first made a breakthrough in 1947, with a special alarm design using an audible metal membrane, vibrating freely so it could emit its full volume. Then, other manufacturers jumped on the bandwagon or remembered, like LeCoultre, that they had already worked with this technology in the 1920s. The Memovox, followed in 1956 by the Memovox Automatic, with Caliber 815 as the first automatic alarm watch, still with a hammer automatic mechanism. In 1969, Omega presented the Memovatic, an alarm watch that could be set exactly to the minute. Some 1.4 million of the A. Schild S.A. manufactured alarm Caliber AS 1475 movement were made up to 1974.

Jaeger-LeCoultre introduced a high-priced alarm watch during the mechanical euphoria of the early 1990s. The Grand Réveil showed the admiring mechanical watch community an alarm that had what it takes. Not only did it combine perpetual calendar, moon phase, and alarm—a hitherto unknown combination—but also, the design of the alarm itself was impressive because to its special Chinese bronze gong and the sound it created, so different from the profane

An alarm for world travelers. Maurice Lacroix offers an alarm along with displays for date and second time zone, making this watch the ideal travel companion.

While alarms were formerly more likely available in the medium price range, this Breguet alarm makes clear that this complication has arrived in the luxury range.

rattling everyone was used to. This Caliber 919 was also offered, as the 918, as a pure alarm with no other complication except date display. Even the most affordable version of this watch, only available in yellow or rose gold or platinum, cost a good $20,000 in 1995, strictly limiting the circle of interested parties. More affordably priced is a remake of the of the Vulcain, available as the Cricket Nautical which can be worn under water. Since an alarm cannot compete with the attractiveness of other complications, some manufacturers decided to present it in combinations. In its Valjoux Caliber 7750-based alarm chronograph, Fortis has manufactured the first-ever combination of both complications; Breguet and Blancpain created, in their Le Réveil du Tsar and Léman Réveil GMT models, an interesting, sensible alliance of alarm and GMT watch.

What traveler wouldn't require a reliable alarm to free them from the annoying hotel wake-up service? Fortunately, the self-winding Blancpain Caliber 1241 winds both watch and alarm movement. In the first manual-winding alarm movements, alarm and watch movement had to be powered separately, using two crowns. Later on came designs in which the watch movement was wound automatically, but the alarm bell still had to be spring-loaded by hand. Nowadays, thanks to self-winding, there is always energy for the alarm; you only require, apart from the alarm indicator, the means to turn it on or off, so you won't be disturbed by a ringing alarm at some inappropriate moment. Should you nevertheless fail to do this, and the alarm sounds inadvertently, you can be sure that any such disturbance will be greeted with sympathy and tolerance by those around you. Thanks to mechanics.

More than a simple alarm. The Sonata by Ulysse Nardin, with second time zone and count-down indicator for the alarm signal.

COUNT DOWN
ALARM SETTING
OFF ON
Sonata
ULYSSE NARDIN
DUAL TIME
SWISS MADE
N° 172

A. LANGE & SÖHNE
GLASHÜTTE I/SA
25
AUF
AB
GMT
MADE IN GERMANY
DOPPELFEDERHAUS
DUBAI
KARACHI
DACCA
BANGKOK
HONG KONG
TOKYO
SYDNEY
NOUMEA
AUCKLAND
MIDWAY
TAHITI
ANCHORAGE
L.ANGELES
DENVER
MEXICO
NEW YORK
CARACAS
R.JANEIRO
S.GEORGIA
AZORES
LONDON
BERLIN
CAIRO
MOSCOW

The World Time Watch—Time Zones in GMT and/or UTC

The impact of rapid industrialization and rising rail traffic from the mid-19th century, made it necessary to newly define distinct time zones, nationally and internationally. In 1875, the globe was divided into 24 time zones. This was initiated by an engineer named Sandford Fleming, who proposed dividing the earth's 360 degrees of longitude by the number of hours in a day; in each respective one-hour gradation, in the range of 15 degrees longitude, the same time would be in effect. The small English town of Greenwich, today part of London, was selected for the prime meridian. GMT means nothing other than Greenwich Mean Time. In 1893, Germany set a unified national time. The International Date Line is located at the anti-meridian, the 180th longitude, roughly between Samoa and Auckland. The UTC, or the Coordinated Universal Time, was introduced in 1919, to allow global comprehension of the exact time designation, by using a four-digit number to make it safe from error.

This applied to both the way the time is displayed and how it was expressed in speech. In the English-speaking world especially, with its additional specification of before noon (a.m. = ante meridian) and after noon (p.m. = post meridian), difficulties in communication were inevitable. Therefore, the ambiguous "2:35 a.m.," becomes spoken as "zero two thirty-five" and written as 0235 in UTC—totally clear.

Wristwatches with world time display have been on the market since sometime in the 1930s. Any watch with a second separately adjustable hour hand, theoretically has the potential to be set for a second time zone. However, you would only be able to know at the particular location itself, whether it is day or night there. As a result, a time zone watch requires a day/night or 24-hour display. This is often set on the bezel, or is imprinted as a 24-hour ring outside or inside on the dial. The originally non-adjustable GMT hand circles the dial once every 24 hours. Of course, the 24-hour display can also use a small sub-dial,

World time at its finest. The watch also has a power reserve indicator, Big Date, and day/night display, important with only a 12-hour scale.

or the watch can have two 24-hour displays, something that can be confusing for those who are accustomed to conventional watches.

When we come to a watch that depicts multiple world time zones, we are dealing with a world time watch. The city names that stand for the time zones are usually imprinted on the bezel or watch dial. Any phone or laptop has a similar function. The Jaeger-LeCoultre Géographique displays world time on a second dial with day/night indicator; a less expensive but not as sophisticated version is offered by the Oris XXL Worldtimer. This design can use up to four sub-dials. Other models identify the time zones by cities, displayed with the corresponding time of day in windows. These can be adjusted either by using the crown or, more quickly and practically, using the one or two push-pieces provided. Two push-pieces make it possible to switch forward and back.

Blancpain offers an attractive solution in its Time Zone, reminiscent of a moon phase watch. In addition to a second time display under the twelve, there is a day/night indicator at the nine in a semicircle shape, which shows either moon or sun, leaving no ambiguity about whether it is day or night. Some GMT watches that are still popular include the Rolex GMT-Master II and GMT-Master models. Ulysse Nardin has a really unusual time zone watch with perpetual calendar and a time zone display easy to set forwards or back using a push-piece.

The Ulysse Nardin GMT offers a second time zone in a digital display at the nine (left). What looks like a moon phase is actually the day/night display on this Blancpain time zone watch (right).

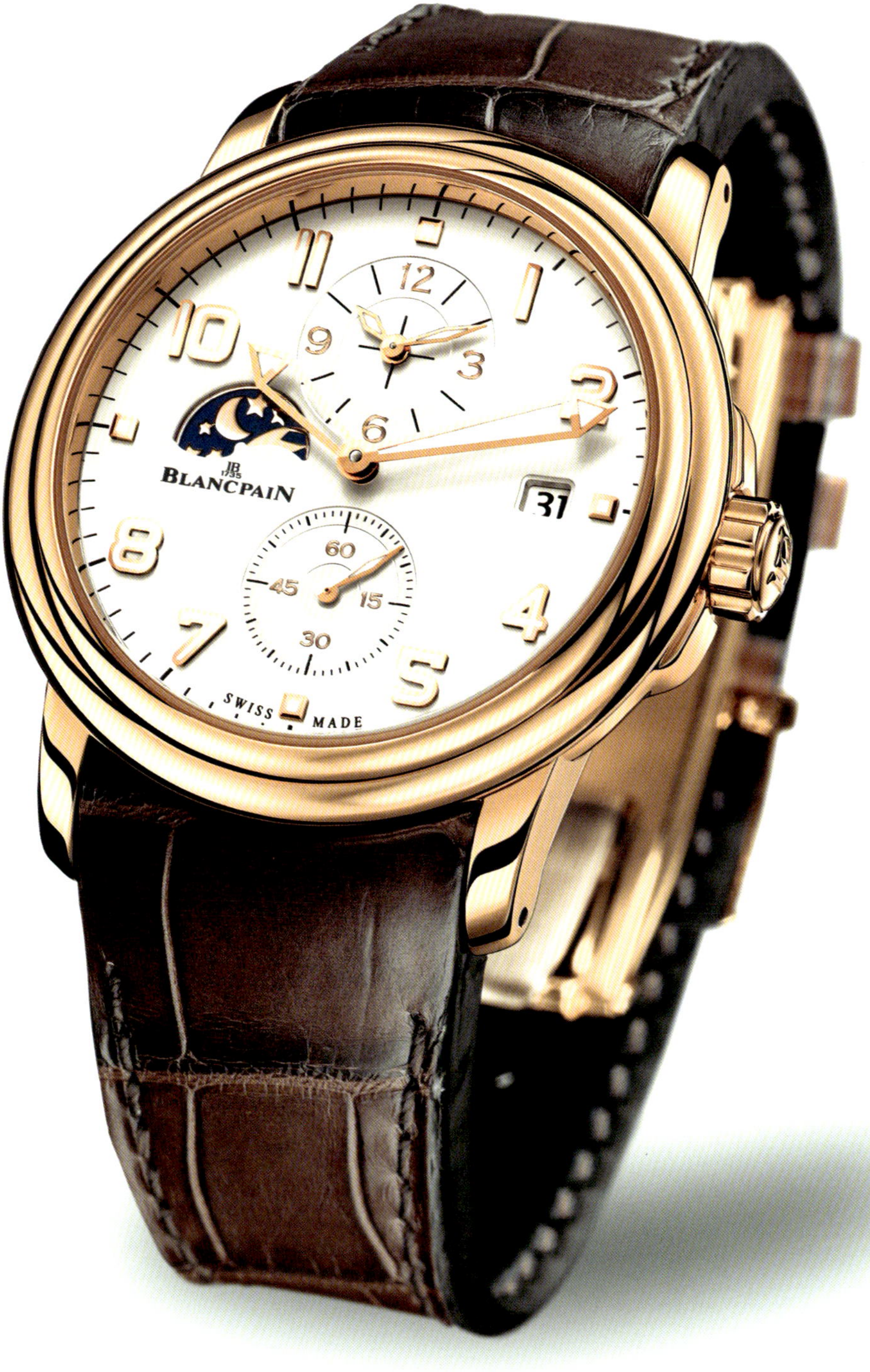
BLANCPAIN
SWISS MADE

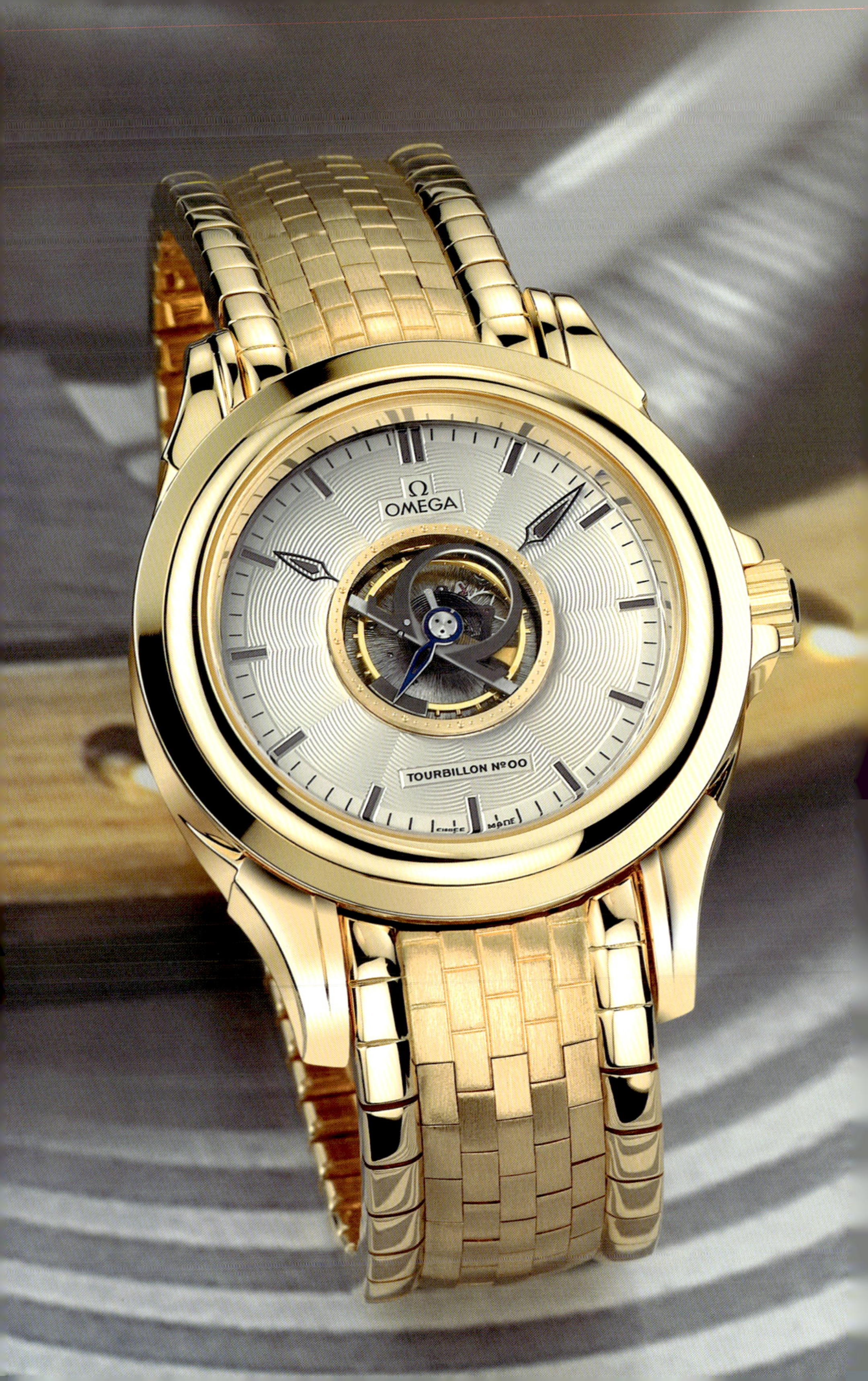
Ω
OMEGA
TOURBILLON Nº00

The Tourbillon—Watchmaking's French Revolution

There are many factors that affect the precision of a mechanical watch, including the earth's gravity. This is especially so when the watch is constantly carried in a pants or vest pocket. Every small offset of the balance has a negative effect on the watch's accuracy rate. It was Abraham-Louis Breguet (1747–1823), likely around the year 1795, who developed a technical solution to overcome this dilemma. The *tourbillon* (Eng.: whirlwind) is an invention in which the anchor, escape wheel, and balance are mounted in a delicate design on the shaft of the second wheel. The entire ensemble can rotate once a minute and thus compensate for positional errors and center-of-gravity irregularities. It was only six years after his invention that Breguet applied for a patent, issued in 1801. Breguet wrote to the responsible Minister: "I succeeded in this invention by preventing, by means of compensation, the errors caused by the positioning of the movement and shifting of the center of gravity, as well as distributing the friction evenly on all parts of the pivots in the said work, even if the oil thickens. Further, I remedied other errors that more or less affect the movement's precision, in a way that by far exceeds the current state of our knowledge."

In most designs, the cage rotates 360° in the course of a minute, making it possible to mount the second hand directly on the cage axis. There were also designs that had it turn every four or six minutes on its own axis, but this brought no discernible advantage.

In the history of watchmaking, few have mastered this challenging design. In Germany it was primarily Alfred Helwig and A. Lange & Söhne, who put themselves in the spotlight with outstanding tourbillion constructions. The most remarkable version is the "flying tourbillon" developed by Helwig, which is

An Omega Central Tourbillion. The time display hands are etched on transparent crystal discs, which move from the case rim. The second hand is part of the central tourbillon, which is self-winding.

The Blancpain Carousel shown above has a special design with the balance rotating like its namesake..

cantilevered or attached on only one side, giving a free view into the tourbillion.

Due to the high proportion of handcraft work they require, tourbillion watches are exorbitantly expensive.

Today, watch connoisseurs wear their tourbillions on their wrists. Actually, the tourbillion is unnecessary here, because the arm movement automatically compensates for all positional and center-of-gravity irregularities. However, it is used in wristwatch movements for the sake of even greater precision. At the Observatory competitions—the Formula 1 of the watchmakers' guild—in terms of maximum accuracy, made much use of in the promotional material published subsequently, Patek Philippe entered a tourbillion, the Caliber 34 T, on the starting line between 1958-1966. Despite the elaborate design, compared to its competitors, the 34 T tourbillon won a double victory in 1962, and that was just as a prototype that customers were unable to purchase. In these competitions, the artistic quality of the respective regulator was often more important than its construction. Nevertheless, in 1950, Omega also entered a tourbillon in the race, which was also victorious.

Audemars Piguet offered a flat self-winding tourbillon watch for sale only in 1986, amidst the mechanical watch renaissance. To give the purchaser not only a watch, but also the most desirable talking piece possible, the rotating cage was made visible by an aperture in the dial. This makes it possible for your counterparts to see at a glance just what a treasure you are wearing on your arm. Despite the computer-controlled, electro-erosion machines that make low-cost manufacture—compared to hand-crafting—of the small parts possible, the tourbillon is still the most expensive complication there is. It is an anachronism, but at the same

Not every manufacturer skeletonizes the dial where the tourbillion is positioned, as this subtle 10-day manual winding chronometer-certificated tourbillon in platinum case demonstrates. In the Reference 5101, Patek Philippe has taken understatement to the extreme edge.

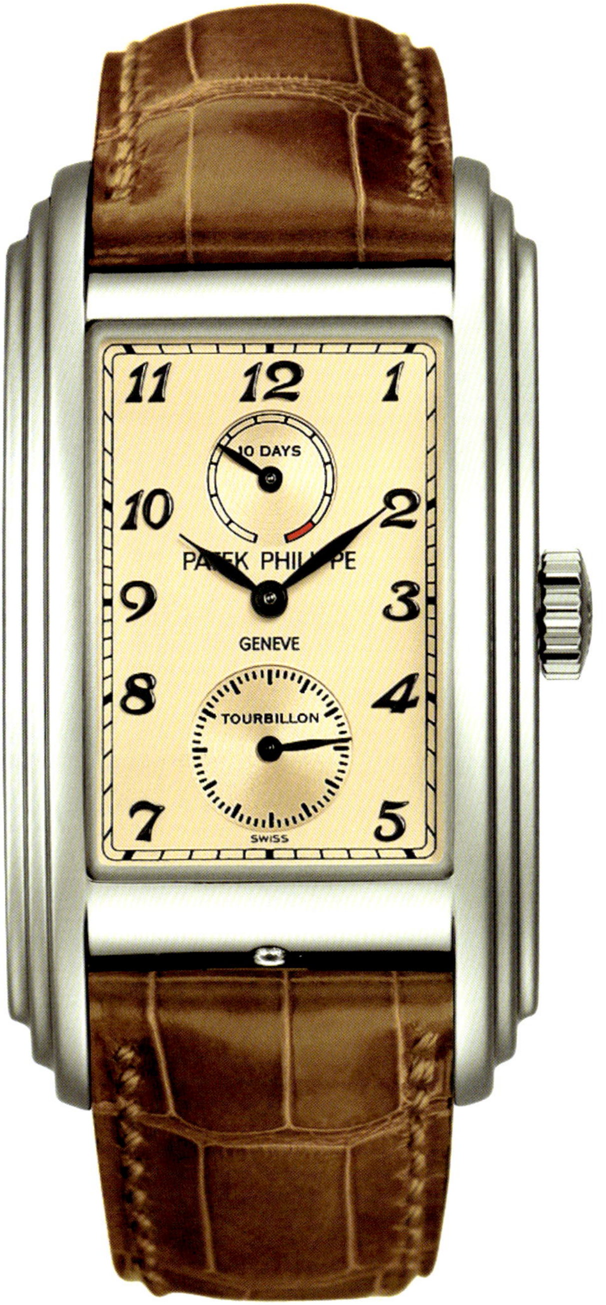

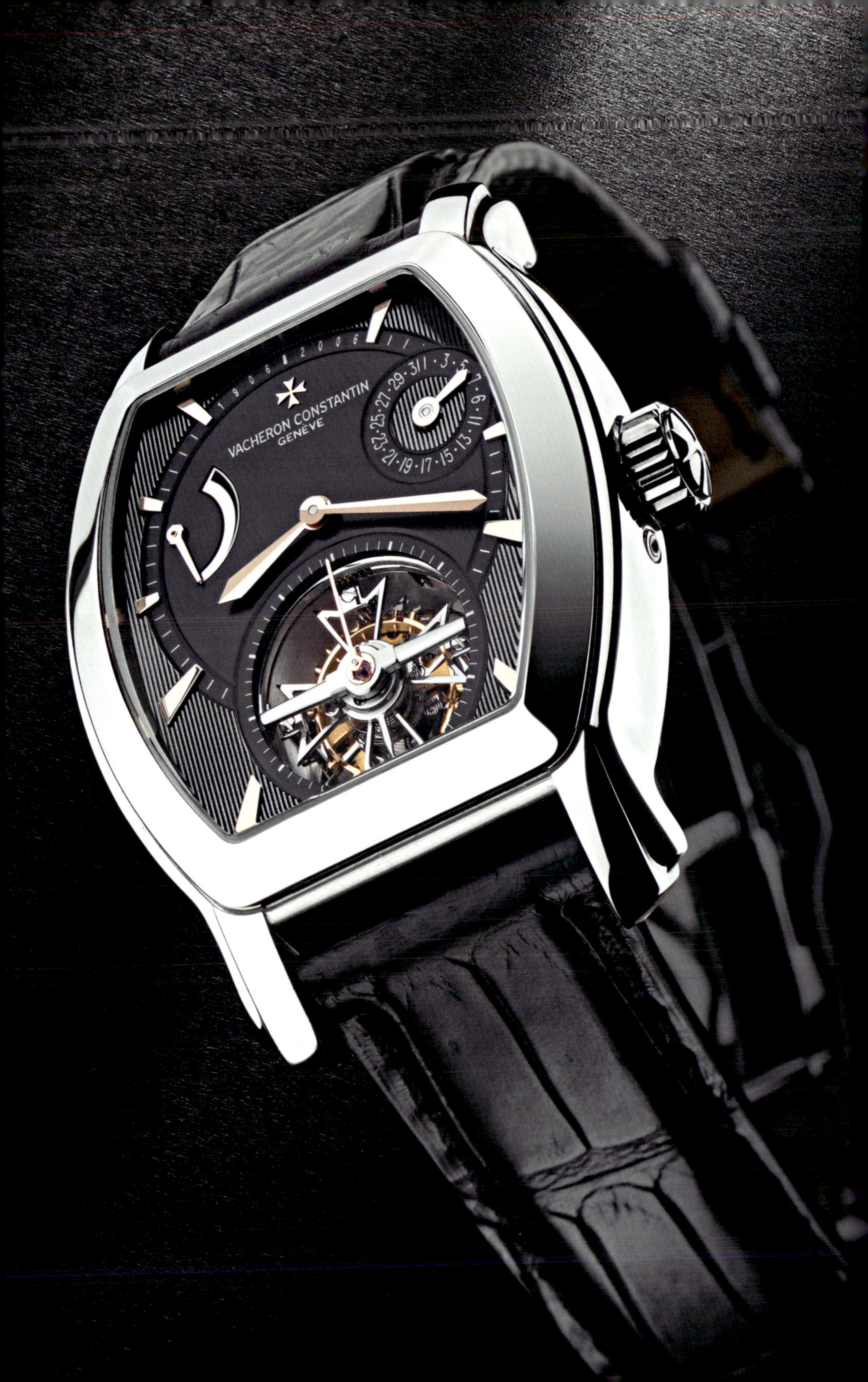
VACHERON CONSTANTIN
GENÈVE

The Maltese Cross, trademark of Vacheron Constantin, adorns the tourbillon. The cage rotates once per minute and thus can function as the seconds display.

Cabaret Tourbillon by A. Lange & Söhne in a platinum case. This manual winding movement is the first tourbillon with stop seconds; this means it can be adjusted precisely, something previously not possible for a tourbillon.

The tourbillon turning on itself is now more of a visual attraction than way to increase the watch's accuracy. This is why most watchmakers skeletonize the dials at their tourbillon models.

time a fascinating design, much like the mechanical wristwatch itself. Perhaps for this reason, it is a special symbol of love for the mechanical watch.

Some twenty years after the introduction of the first tourbillon in a wristwatch for customers to purchase, the watch industry was pleasing its public with a wide range of wristwatches with this complication. There are also wristwatches that feature the complete movement turning on its own axis, corresponding design-wise to the tourbillion concept. Vincent Calabrese offers a transparent watch that displays the moving rotating in the center, as if held by an invisible hand. Ulysse Nardin's Freak model is another watch in which the entire movement rotates. Once such a play instinct is unleashed, it knows no boundaries. Jaeger-LeCoultre's Gyrotourbillon I even features a design with a spherical tourbillon rotating on three axes. The tourbillon not only rotates on its own axis, but also sideways. Although this delicate structure comprises 90 individual parts, it nevertheless tips the scale by a mere 0.336 grams. Most tourbillions are manual-winding watches. Only a few, such as Blancpain, IWC, and Girard-Perregaux, with its Three Bridge Tourbillon, offer an automatic version.

IWC includes a tourbillon in its Portuguese series. The Tourbillon Mystère consists of 79 individual parts and appears to float against its black background. The manually wound movement has a power reserve of seven days.

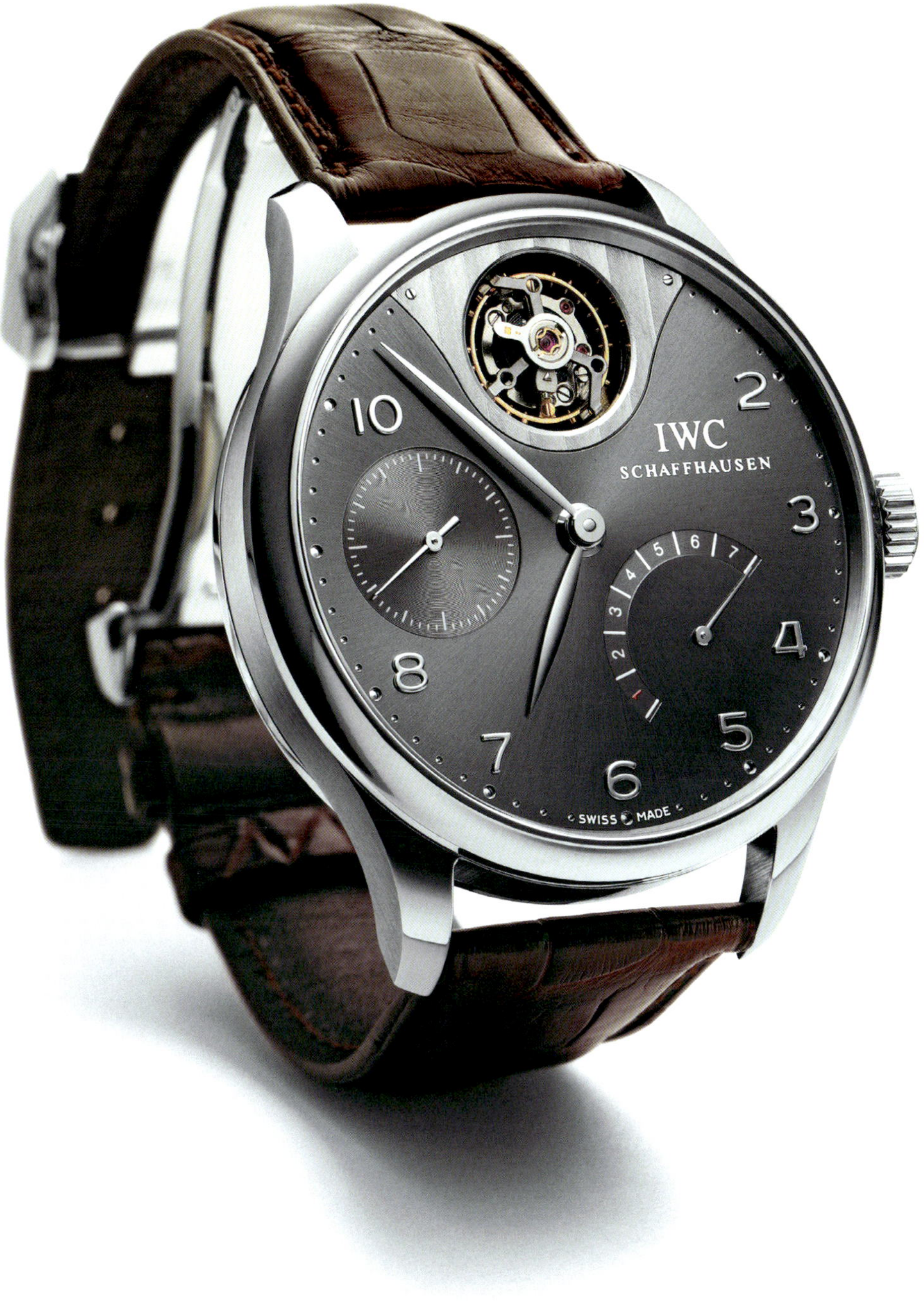
IWC
SCHAFFHAUSEN
SWISS MADE

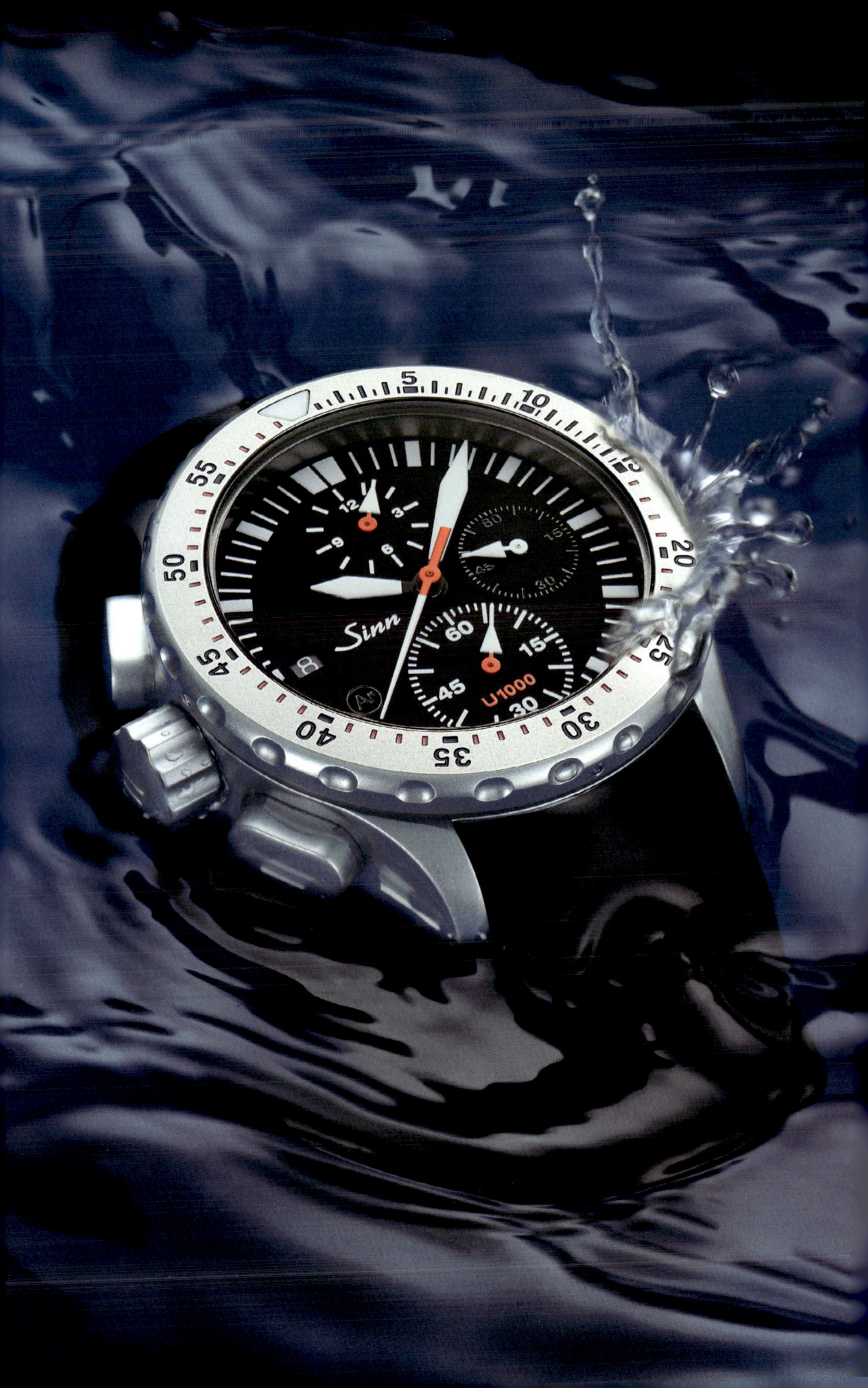
Sinn
U1000

The Diving Watch—The Only Really Robust Everyday Watch

A diving watch is not an independent complication. It is much more the primary product of a decades-long effort to create a waterproof watch. From the very beginnings of mechanical watches, water has been a natural enemy for this mechanical marvel, which was initially made of iron. It was not until the 1920s that a case structure appeared on the market that made it possible to hermetically seal the movement against the external elements. The Rolex company, registered as a brand name in 1908, was a trailblazer in this development. Rolex is an artificial name, derived from *"horlogerie exquise."* The company founder was the German Hans Wilsdorf from Kulmbach, who had set himself the goal of manufacturing a precise wristwatch that could compete with the pocket watches still commonly in use.

Already by 1914, he was able to obtain a Class A certificate of accuracy for a wristwatch movement from the Kew Observatory in England. Yet beyond mechanical durability and excellent precision, Wilsdorf wanted above all to make a watch that would be absolutely watertight. This goal was achieved by sealing and screwing down the case parts against each other, a special screw-down crown design and a positive-locking crystal. A name was quickly found for this work of art: Oyster. As early as 1926, Rolex applied for patents in England and Switzerland. The first Oyster was approved in 1926 and presented to an admiring public.

But Wilsdorf was more than a watch technician and maker. He was also a gifted press and marketing specialist, a craft, which at that time operated under the label "propaganda." With a Rolex Oyster on her arm, the young English typist Mercedes Gleitze swam the English Channel on Oct. 27, 1927, making it clear to all the world that the waterproof watch breakthrough had been made. This triumph was announced in

A diving chronograph by the Frankfurt-based Sinn company, which is known for its reliable daily-wear watches. The uni-directional rotating minute ring is used to set the dive time. Crown and push-pieces are mounted on the side away from the back of the hand, to increase wearing comfort and reliability.

Diving watch with integrated depth gauge by Jaeger-LeCoultre; essentially a gimmick, because every diver always takes a separate, large depth gauge along.

a full-page ad on the front page of the *Daily Mail*. In more publicity stunts, Rolex had its concessionaires install small aquariums in their display windows, where astonished window shoppers could see goldfish swimming around the watches. Today, the DIN (*Deutsche Industrie Norm* or German Industrial Norm) standard regulates the difference between a waterproof watch and a diving watch. According to DIN 8310, a watch can be called water-resistant if it withstands immersion at a depth of one meter for 30 minutes. For diving watches, DIN 8306 specifies, among other things, that the watch must remain undamaged when held at the specified maximum water depth for two hours as well as immersion three meters deep for three hours.

One of the outstanding diving watches of the postwar period was again a Rolex. The Submariner, first presented in 1955, was initially approved to a depth of 100 meters and had a rotating bezel, which made it possible to adjust it to the duration of the dive time at the start. Of course, such a watch would also have a screw-down crown. The Rolex Submariner was sold for what is, from today's perspective, the bargain price of $150. Twenty years later, the price had doubled. The inscription on the dial ensured, "Oyster Perpetual Date, Superlative Chronometer, officially certified, 660 ft = 200 m." Inside ticks the automatic movement Caliber 3131 with 21,600 vibrations per hour, a Rolex in-house development, which also served as the basic caliber for the other Oyster models.

The power reserve is 48 hours. The Oyster case, developed by Rolex founder Hans Wilsdorf and patented in 1926, laid the basis for the totally waterproof wristwatch. Expeditions, extreme athletes, and adventurers made this model famous, and soon this

The minute hand is a crucial element of any diving watch, essentially making a regulator dial a must. Oris did this very well.

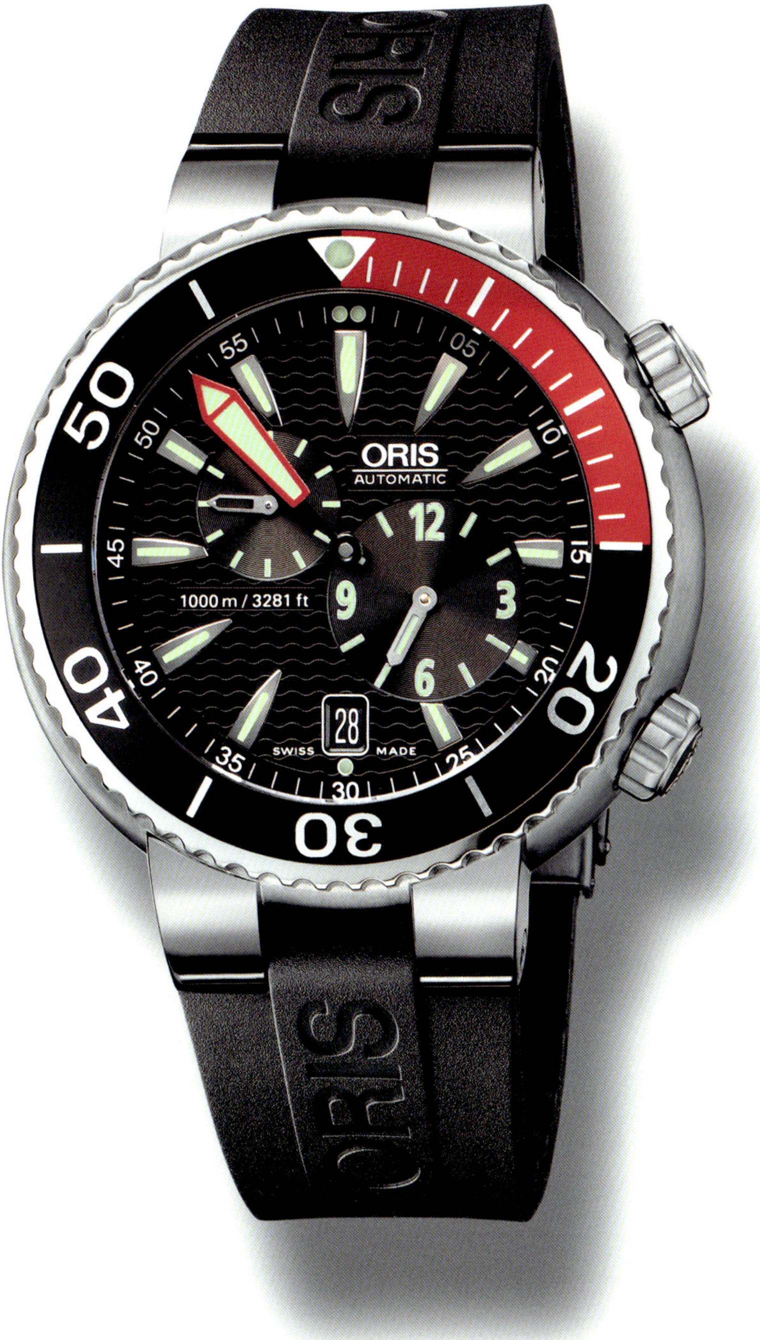
ORIS
AUTOMATIC
1000 m / 3281 ft
SWISS MADE
28

Different styles of IWC diver watches. Chronograph and integrated depth gauge complement the usual diving watch features.

watch species mutated into a showy prestige object. Apart from time, the watch also shows the date through a magnifying glass—the Cyclops—at the three. Unfortunately, this model had no quickset date, so that, if the watch has not been worn and the date has to be re-set, an endless dogfight began. The new models also have a sapphire crystal instead of the plexiglass then usual, a quickset date, and a more advanced Caliber 3133 movement with 28,000 vibrations per hour. The uni-directional rotating bezel lets you make the time adjustment so important for a diver. Luminous Tritium dots on the dial and hands also outlined with this material let you read the time even in complete darkness. True to Rolex's advertising promise, "If you can take it, so can your watch," the Submariner can be worn anywhere, including in the sauna.

While the oxygen rebreathers which were mainly used by military divers would only allow dives to small depths up to 15 meters, this situation changed with the invention of the aqualung by Jacques Cousteau. In 1953, Blancpain presented the Fifty Fathoms, a watch which, as the name suggests, can withstand water pressure down to 91.45 meters. The manufacturer rapidly increased the possible depth range to 200 and even 300 meters.

For such depths, you must know that free divers using compressed air equipment can safely dive to depths of 60 meters at most. Greater depths require a helium-oxygen mixture. Offshore divers, who work under these conditions, will remain in pressurized cabins after a dive, which are also filled with this gas mixture. Because helium diffuses through glass, mechanical watches worn at such extreme depths must be equipped with a helium escape valve, or the crystal will explode as the external pressure slackens.

Like all other mechanical watches, traditional diving watches are also an anachronism, that has long since been replaced by the dive computer. Divers now also wear a mechanical watch only as a safety reserve. The traditional dive watch is a three-hand watch with rotating bezel—today they usually rotate only counterclockwise, although that was not the case for the first models. A classic diving watch did not include complications such as a chronograph function. The manufacturers were happy not to load down the watch with additional complications. Anyone looking for a rugged everyday watch, which won't be affected by a visit to the sauna, including that final jump into the cold water pool, should absolutely decide on a diving watch.

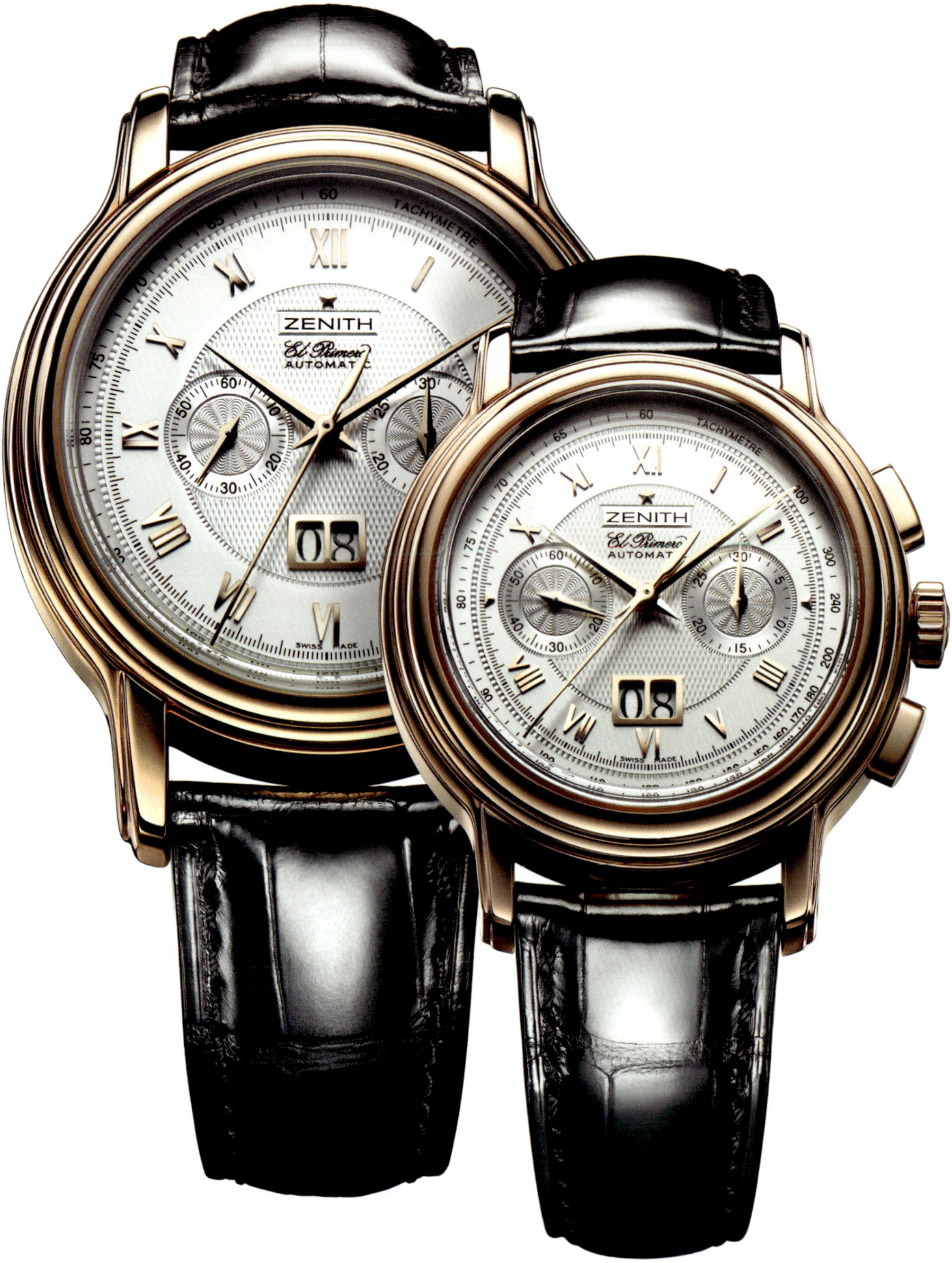
ZENITH
TACHYMETRE

The Big Date—The Date at Center Stage

Everything that takes a watch beyond the three-hand design is a complication, differentiated between larger (such as the perpetual calendar or tourbillon) and smaller complications. The Big Date is one of the few small complications specially presented by A. Lange & Söhne as a contemporary innovation some time after the mechanical wristwatch revival. For people with sight limitations, watch complications are something that can either be read with difficulty or only with eyeglasses or a magnifier. Anyone who wants to read the year or month on an IWC Novecento or a Blancpain Perpetual Calendar, needs very good eyes, which older people usually do not have. What could be more important than to make the display which is the most important, next to the time—the date—large enough so that it is easy to read without glasses?

At first, watch manufacturers tried to do this using a magnifier glued on the crystal—a functional but not very aesthetic solution. The big date was one of the key design elements from the A. Lange & Söhne company, newly founded in 1990. Already in 1994, Lange 1 amazed the watch world with a date display five times larger than the usual design. It is particularly evident in the rectangular Cabaret model, in which the date ring, with imprinted numerals as generally used, would encompass the whole movement. The Lange patent is based on a two-disc mechanism. One ring-shaped disc is printed with the numbers from 0 to 9, above it is set the cruciform month disc with the numbers 1 to 3 and a blank space. Both discs switch forward in increments and automatically continuously show the 1st to the 31st. The watch will not switch to 32, but halts at 1, and the tens disc is switched into the blank space. This is run by a complex mechanism, which works using program wheels and a precise toothing pattern. As long as no perpetual calendar is included, however, the date must be switched forward by hand in the shorter months. The big date gained such great market acceptance that other

The big date in matching outfits. The fast beat from Zenith with 36,000 vph is still one of the most beautiful chronograph movements

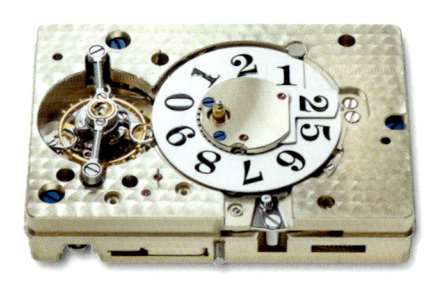

The new A. Lange & Söhne tourbillon also has a Big Date display.

manufacturers tried to offer watches with big date display, always in search of new technical solutions that would enable them to bypass the Lange patent. Just one year after the launch of the Lange Big Date, its local competitor, Glashütte Original, also presented a big date display. This took its own path, as is obvious from the fact that the date discs are set at one level, making the divider bar between the two digits obsolete. The conceptual inspiration for the Lange big date was the 5-minute clock designed by court watchmaker and castle clock-tower dweller Johann Christian Friedrich Gutkaes for the Dresden Semperoper, which also has a divider bar. The solution selected by Glashütte Original, using an inner ring for the tens and an outer one for the single digits, allowed a date display about 3.5 times larger.

Depraz Dubois, a Swiss company specialized in cadrature or under-dial work, developed a module for the ETA 2892-A2, which is offered by various manufacturers in different forms with other complications, such as a chronograph. This uses two superimposed date discs. The tens disc has windows, which allows you to see the single-digit disc. Another solution is provided by the watchmaker Comor's Grande Date. In this watch, equipped with a Venus Caliber 221, the discs imprinted with the single numerals and the one with the tens numerals lie opposite each other. The Ulysse Nardin Ludwig perpetual calendar, designed by Ludwig Öchslin, features a large double-digit date display. Besides the perpetual calendar by A. Lange & Söhne and Glashütte Original's Caliber 39, this is the only perpetual that combines both of these complications.

The Big Date originated in Saxony, and anyone interested in this particular date style is recommended, depending on their taste and budget, to choose a model made in Saxony.

No other watch big date display is as amazing as the Cabaret's. You can't help wondering how it was contrived, because a date ring from 1 to 31 should extend beyond the circumference and diameter of the watch.

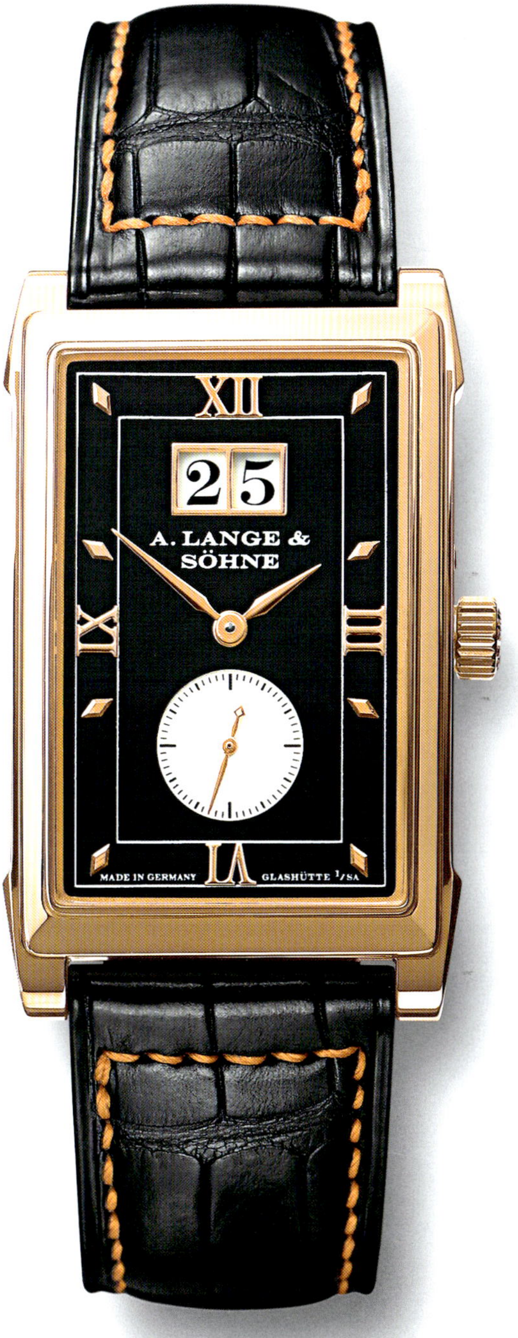
25
A. LANGE &
SÖHNE
MADE IN GERMANY
GLASHÜTTE I/SA

VACHERON CONSTANTIN
GENÈVE
SWISS
MADE
PERPETUAL
CALENDAR
LEAP YEAR
1ST
2ND
3RD
4TH
SAT
SUN
MON
TUE
WED
THU
FRI

The Grande Complication—Nothing Less than What's Technically Doable

Watch prices can reach dizzying heights. A watch costing the same as a single family home is something irritating for most people today; indeed, some even find it morally reprehensible. Is it legitimate for someone to adorn themselves with such precious items? Such a tradition lives on in the Grande Complication, namely, that of exorbitantly expensive watches. In the national archives, there are invoices and notes from watchmakers, showing that in 1500, one watch cost more than 500 Rhenish guilders. For comparison: at that time, Albrecht Dürer acquired for himself and his wife a five-story house in Nuremberg at the cost of 275 Rhenish guilders. Even for a simple watch, Peter Henlein, a contemporary of Dürer from the city of Nuremberg, still charged 57 guilders. Today, the price of a Grande Complication ranges from $200,000 up to $1,200,000. The Grande Complication pocket watch unites all major complications. Originally, these were the perpetual calendar, minute repeater, rattrapante chronograph, and moon phase. When Omega began industrialized wristwatch manufacturing in 1900, it was a development many watchmakers viewed more than skeptically. No one would have thought it possible to miniaturize a Grande Complication to the extent that it would fit on your wrist. Calendar watches only emerged in the 1920s. It took another twenty years before perpetual calendar watches were introduced as ready for serial production. Combining all the important complications to create the Grande Complication was first done in the 1950s by Patek Philippe. Therefore, it took more than half a century until this watch design found its way to the arms of wristwatch lovers. The Reference 1518, a perpetual calendar chronograph by Geneva-based Patek Philippe, can be genuinely considered the precursor of the Grande Complication wristwatch.

Of course, there always were and are people who want something special. These include the owner of a watch with an old

Perpetual calendar and minute repeater are features of the Vacheron Constantin Malte. The manually winding movement with a height of only 4.9 mm runs at a moderate 18000 vph.

Chiming mechanism and chronograph are two preferred complications in a Grande Complication, making it necessary to significantly increase the height. As seen in this Zenith Grande Class Traveller, which also features two time zone displays.

Piguet repeater movement, which was refashioned by Paul Gerber and Frank Muller into one of the most complicated watches ever made. The watch, comprising 1,037 movement parts, features a tourbillon, perpetual calendar with moon phase, and a 24-hour display, as well as minute repeater, Sonnerie, flyback chronograph with splint-seconds, a power reserve display for the movement, and chiming mechanism, and finally, a room-temperature thermometer.

With its Sky Moon Tourbillon 2001, Patek Philippe achieved a masterpiece of design for the Grande Complication. Tourbillon, perpetual calendar, moon phase hands, and minute repeater are enhanced by the depiction of the night sky and rotation of the stars. The features go on to include the displays for sidereal time and angular motion and phases of the moon. This marvel comprises a total of 686 movement parts.

Blancpain and IWC also offer two Grande Complications, the 1735 and the Il Destriero Scafusia—which contain almost 750 parts. These two watches feature tourbillon, split-seconds chronograph, perpetual calendar with moon phase, and minute repeater. The IWC is manual-winding; the Blancpain watch an automatic.

Even someone who prefers a metal bracelet and a watch overall more for sportswear, doesn't have to rule out selecting a Grande Complication. Audemars Piguet offers the Royal Oak, a platinum sports watch which also has the complications noted above. Of course, the price of this watch is equivalent to that of a family home. Accordingly, such watches are made only in extremely small numbers, or are even custom made.

Who buys these watches? Not infrequently, very ordinary people like a Parisian publisher, who takes the metro every morning to his publishing house, wearing his $900,000 Blancpain 1735 on his arm.

The Jaeger-LeCoultre Reverso Grande Complication à Triptyque is a complication watch, which skillfully makes use of its reversible case construction. The perpetual calendar and moon phase are in the base which holds the case.

HORIZON NORD
EST
OUEST
LEVER
COUCHER
EQUATION DU TEMPS
SWISS MADE
A TRIPTYQUE
SAM DIM
JEU MER MAR
NOV JAN MAR
SEP JUIL MAI
INSTANTANE

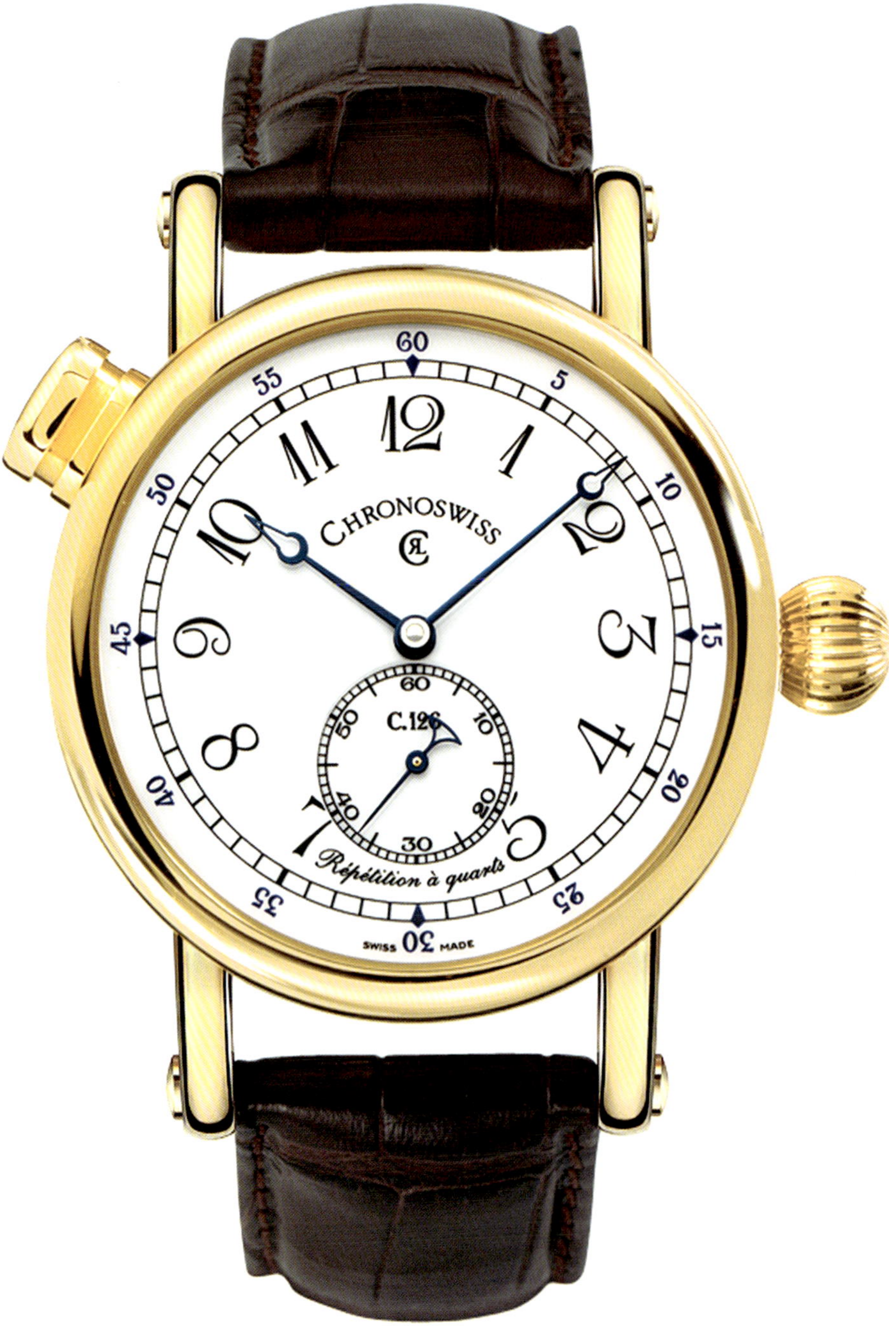
CHRONOSWISS
C.126
Répétition à quarts
SWISS MADE

The Repeater—Gentle Tones from a Watch

Next to the chronograph, the repeater watch, which proclaims the time on demand by a delicate chime, is the only complication that lets the owner actively intervene in the watch movement. Since a repeater chiming mechanism is an extremely expensive additional feature, sending the watch price upwards accordingly, this is a complication rarely encountered and rarely made in a pure form. More often, it comes as the icing on the cake in a Grande Complication. The origin of this complication harks back to the much darker past times, when there were no electric lights and no luminous dials. If its owner was either lying in bed at home, or in a hotel bed while travelling, and wanted to know the time during the night, they would press the slide of his repeater pocket watch. In the silence of the night, a fine tone would sound, reminiscent of a cricket's chirping, telling the exact time. Depending on your budget, the time of day would be sounded more or less accurately. The less expensive pieces would sound the quarter-hour, eighth of an hour, or five minutes, or, according to the owner's wealth, a minute repeater would sound. This would then strike the time exactly: hours, quarter-hours, and preceding minutes, on two sounding springs or gongs, which chimed either alone or in a duet. The hours were chimed in a darker tone, the quarter-hours as a duet, and the minutes in a brighter tone. The power for this chiming was generated by a push-piece or slide. If not pushed with enough force, the watch would strike the wrong time. Only the best designs had an "all-or-nothing" safeguard to prevent the owner getting the wrong time because they had not operated the watch correctly. The all-or-nothing would prevent the watch from sounding at all. There were already quarter-hour repeater watches by the 17th century; those with the more complex cadrature for a minute repeater appeared first in the 18th century.

For a long time, the repeater was a forgotten complication, made obsolete by

From Chronoswiss—only striking the full and quarter hours. A push-piece at the ten starts the mechanism.

electric light and luminous dials. The watch companies displayed such models, if at all, only as individual showpieces, as Piaget did. In 1955, this company presented a unique Grande Sonnerie, based on a 1910 movement design. It was only with the mechanical watch renaissance that this forgotten complication returned to the wrists of our well-heeled contemporaries. Blancpain was the first manufacturer to offer this complication as a solitaire in a wristwatch in the mid-1980s.

The tourbillion was powered by a 9-ligne bridge movement with a diameter of 23.5 mm and a height of 6 mm. The 351 movement parts were assembled together in the tiniest space. In this piece, the problems were many times greater than for assembling a pocket watch. The necessary gong was so much smaller. What was wanted was not a shrill sound, but something melodious. This can only be created with appropriately fine and deep tones.

Of course, this chiming mechanism can be enhanced to make a Sonnerie or even a carillon. The latter translates to bell chimes and is always used when a bell movement can play a melody. The Sonnerie is divided into a Petite Sonnerie, which sounds at the full hour, and a Grande Sonnerie, which starts sounding on the quarter-hour—however,

Gongs and hammer can be easily seen; they provide the harmony created by a minute repeater.

At first glance, a simple three-hand watch. However, in 43-mm case of IWC's Portuguese conceals a minute repeater that strikes the hours, quarters and minutes.

only at one pitch. If you want to experience the full euphony, you must use the slide.

A minute repeater with Grande Sonnerie is a watch that might not immediately catch the eye of a layman, but certainly will catch that of a connoisseur. Its appearance is as modest as a normal three-handed watch, and only the slider on the side tells the connoisseur what a horological treasure it represents.

The repeater is a quiet complication. Anyone at a dinner party, who tries to sound their minute repeater amidst the general conversation, will soon learn that it cannot be heard in the murmur of voices.

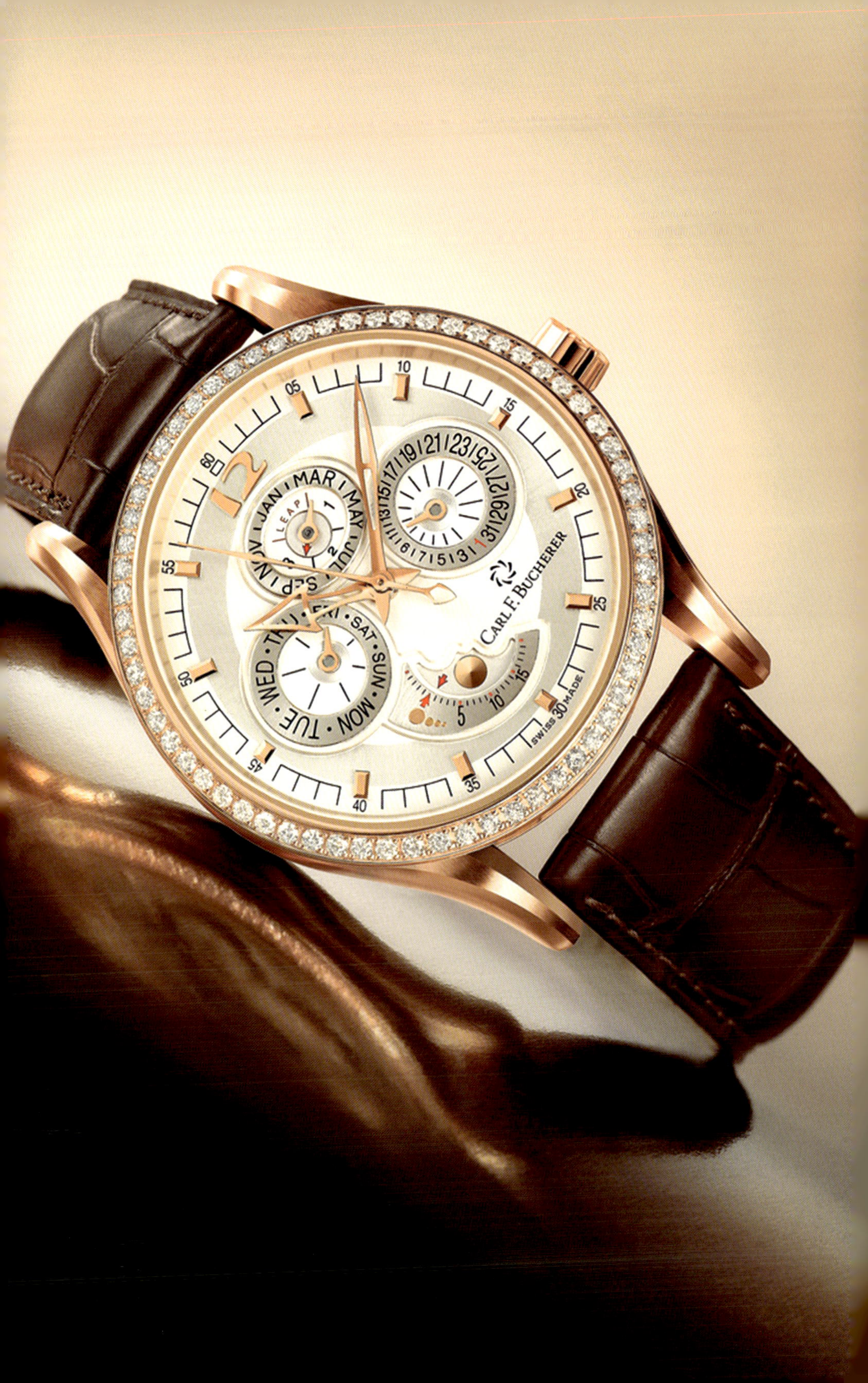
CARL F. BUCHERER
SWISS MADE

The Perpetual Calendar—A Calendar There's No Need to Set

Perpetual calendars now—in contrast to the past—surround us in many shapes, which we often don't even consciously register. Every cell phone and computer, even the simple organizer, provides us with the necessary calendar data. From the 18th until the mid-20th centuries, this information was in no way so easy to obtain. Anyone possessing a perpetual calendar in their pocket watch had a mechanical, precision instrument at hand that not only always told them the exact different lengths of the months, but also noted the sequence of leap years. As early as 1764, English master watchmaker Thomas Mudge presented a pocket watch with perpetual calendar, and by the 19th century all the well-known manufacturers offered this complication to their pampered clientele. However, it was only in 1936 that a Patek Philippe perpetual calendar with retrograde display was also available as a wristwatch.

The renaissance of the mechanical watch, which owed its origin not least to the attractions of the calendar watch with moon phase, also led to a comeback for the perpetual calendar with moon phase display. Reduced from pocket watch to wristwatch dimensions, as many manufacturers have and still manufacture these pieces, the various indicators that a perpetual calendar inevitably includes are, however, difficult to recognize. At a time when the big date was creating a furor in watch design—what could be more obvious than a full-scale re-design of the displays. Not only the Swiss, such as Ulysse Nardin, but also the Saxons did just that, and decided to use window displays throughout instead of hands. This is what was done with the perpetual calendar watch in Glashütte Original's classic Senator model series. On this watch, weekday, month, date, moon phase, and leap year are all displayed in

Carl F. Bucherer's Manero Perpetual Calendar has a moon phase scale as a special feature and is a certified chronometer.

IWC offers a perpetual calendar with moon phase display in a Portuguese case, which shows both southern and northern hemispheres. The watch also has a countdown indicator to show the time remaining until the next full moon.

This perpetual calendar in a stainless steel tonneau case pays homage to IWC designer Klaus Kurt. Special features include the four-digit year display and calendar quickset by the crown.

windows and are much more clearly visible than could be possible by hands. While the month and date display at the ten and two recall the Valjoux Caliber 72C, 88, or Zenith Caliber 410, the diagonal juxtaposition of moon phase at the eight and the big date at the four, gives this dial a distinctive look no other watch has in this form. This watch does display the time as on any normal three-hand watch.

The cadrature for a perpetual calendar is a complex additional mechanism, usually mounted on the watch movement. It is the mechanical realization of the Gregorian calendar, but it doesn't take the loss of three leap-days necessary every 400 years into account. A century year is only a leap year when it is divisible by 400: thus, the years 2100, 2200 and 2300 are not leap years, and the watchmaker must then correct the

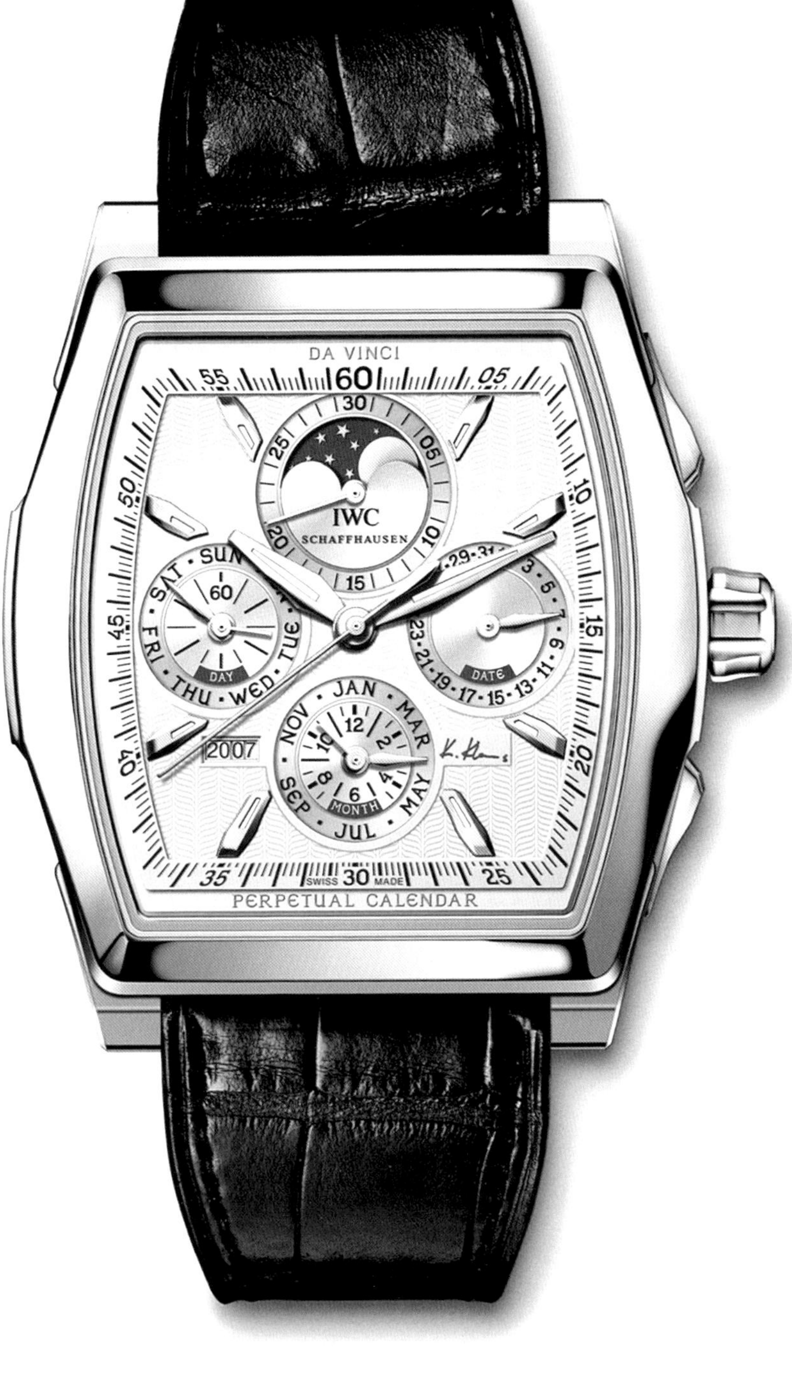
DA VINCI
IWC
SCHAFFHAUSEN
2007
SWISS MADE
PERPETUAL CALENDAR

ZENITH
El Primero
GUILLOCHE MAIN
QUANTIEME PERPETUEL
TOURBILLON

calendar by hand. As such, this is not actually a "genuine" perpetual calendar.

Only perpetual calendars that have all functions controlled by the crown require manual correction by a watchmaker, not those with a corrector push-piece set in the case. Both solutions have their pros and cons. If you control the functions with the crown, you need only set the correct date and the moon phase will be correct. However, once a display jumps, only a watchmaker can help; with a watch with correction push-piece, the owner can take care of it themselves. The information on the correct month length is provided by a month cam with appropriate cutouts.

Most perpetual calendars show the every-four-year shift to a leap year with a hand which covers over the first, second, and third years with the subsequent leap year. There are, however, also year displays, such as made by IWC, that show either the last two numbers or the full four digits for the year. Here, you no longer know whether you are in a leap year or not, but it also doesn't matter, because the calendar manages this automatically. If the display shows the entire date, it requires a century slide, which operates in the incredible ratio to the balance of approximately 6.3 billion to one.

Vacheron Constantin offers a very slim perpetual calendar with chronograph in this manual-wind watch.

Today, watches with perpetual calendar are usually automatics. Again, it was Patek Philippe that launched the first watch with perpetual calendar and automatic movement in 1962. Almost all perpetual calendars can only be adjusted in one direction, and that is towards the future. If you have set it too far ahead, your only option is to leave the watch as is until that date arrives, or send it to the manufacturer. Only the Ludwig perpetual calendar by Ulysse Nardin can be freely switched back and forth. This remarkable watch, which features a completely re-designed calendar mechanism, also displays the GMT time as an additional function.

Perpetual calendar as a fast beat movement, with 36,000 vph and tourbillon. Clearly, the Zenith El Primero also has a chronograph, making it possible to measure the tenths of a second.

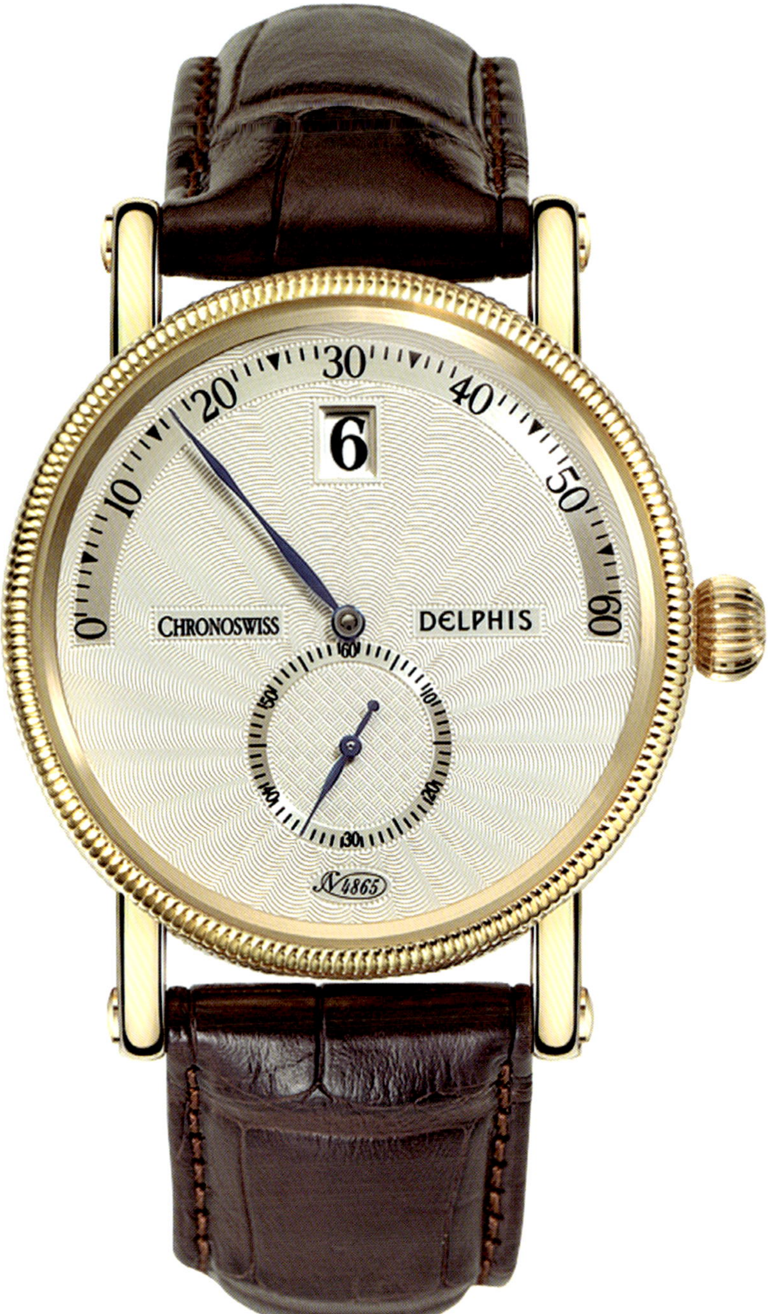

CHRONOSWISS
DELPHIS

The Retrograde Display—Play Instinct with Surprises

Amidst growing enthusiasm for mechanical features, the search began for new technical gadgets. Among these is the retrograde display; retrograde actually means "moving or tending backwards." The Swiss brand Gérald Genta was one of the first to present this complication in the 1990s, still as a unique feature at the time. But, like so many things involving watchmaking, this complication has been known for a long time. There were 17th century pocket watches with this form of display. The famous master watchmaker Breguet made a perpetual calendar watch for French Queen Marie Antoinette, where the date is displayed retrograde. Designing a jaquemart—a moving automaton—which indicates the time by its motions, would be impossible without retrograde display technology. On a Druid, a wristwatch that has a Druid on its dial, his outstretched arms each point to the hours and minutes; after the time has elapsed, they jump back to the starting position. Often, the circular case is divided or quartered for design reasons, when a retrograde display is applied. In the early 20th century, the Records Watch Co. presented the Sector watch, which showed hours and minutes on a 110° arc.

Different systems are used to make it technically possible for the hands to jump back. Often a cam disc is used, combined with a reset spring to move the hand back to the starting position. Another possible system uses an eccentric cam, connected by a scanner to a counter, in turn connected by steel spring to a second counter. Maurice Lacroix uses gear wheels with missing teeth; after it runs through, the spiral spring's force is released to the wheel and thus the hand, jumping it back to the starting position.

But most customers are more interested in what is happening in front of the dial. There are many possibilities, such as moving figures. Gérald Genta has shown us a Mickey Mouse swinging a golf club. Retrograde displays can show not only the date, hours, and minutes, but seconds as well—a

The Chronoswiss Delphis offers three different display styles. The hours are digital, the seconds analog, and the minutes display is retrograde.

The Breguet Classique, offers a hand-guilloched dial and retrograde seconds display, the quickest way to enjoy this complication.

Vacheron Constantin's Bi-Retro Day-Date offers retrograde day and date display. Like many watches from this House, this model also features the Hallmark of Geneva.

complication that really lights up the dial. Some manufacturers will only use one retrograde display, while others, such as Pierre Kunz, use retrograde technology for all three time displays. This style of display works well with other complications. Jaeger-LeCoultre's Reverso Gran Sport Chronograph has a stop watch with the minute counter in a retrograde display. Combining it with a jumping hour is popular, as on the Chronoswiss Delphis. This watch offers the minutes on a retrograde 180° scale and a jumping hour display. Only the second hand moves around its orbit in the usual way. The Maurice Lacroix Masterpiece Double Rétrograde combines a 24-hour display with retrograde date; second time zone and power reserve are displayed retrograde. There are also tourbillon watches by such manufacturers as Glashütte Original or Patek Philippe available with

1 · 3 · 5 · 7 · 9 · 11 · 13 · 15 · 17 · 19 · 21 · 23 · 25 · 27 · 29 · 31
VACHERON CONSTANTIN
GENÈVE
MONDAY
TUESDAY
WEDNESDAY
THU
FRIDAY
SATURDAY
SUNDAY
SWISS
MADE

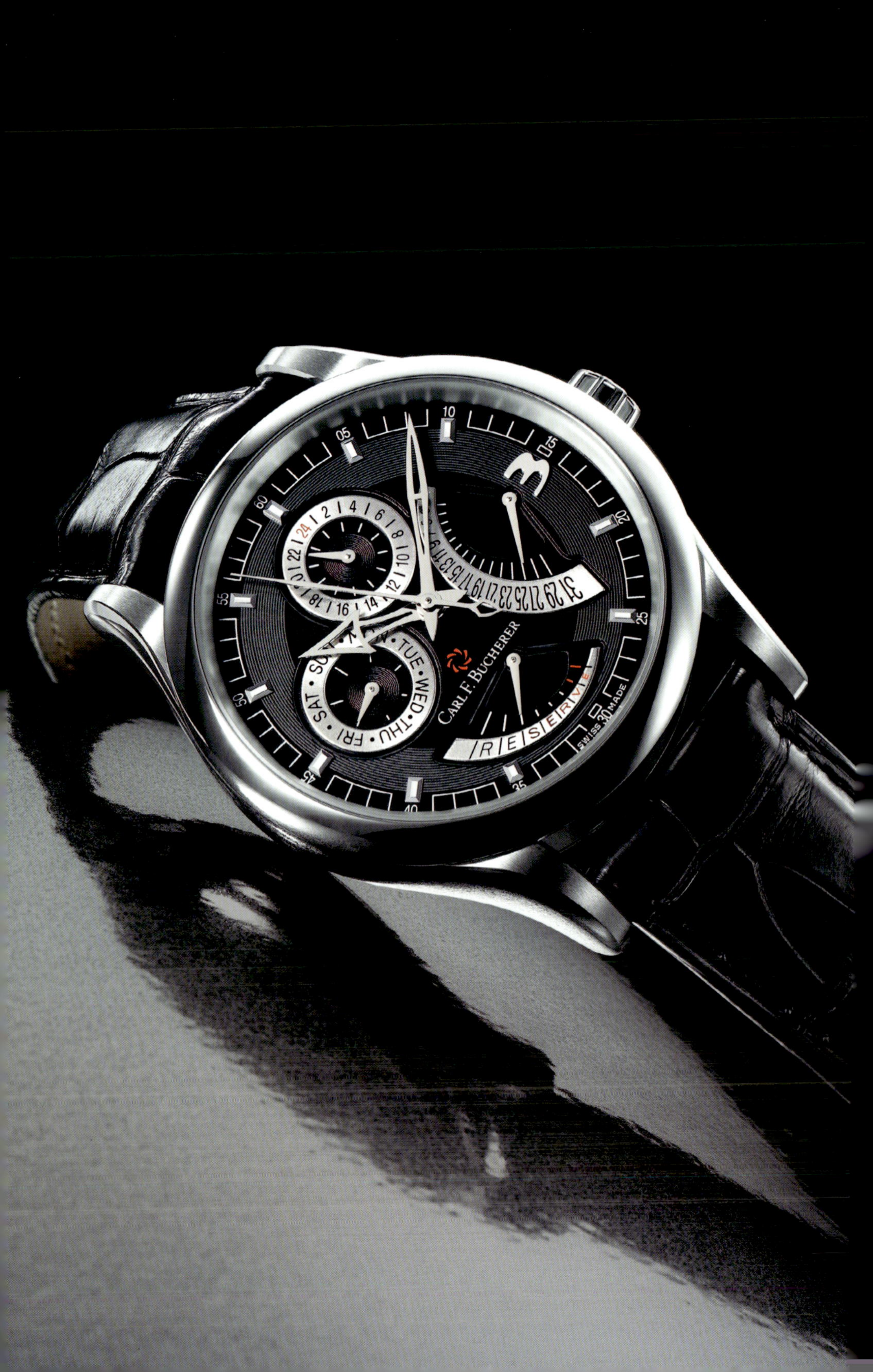
CARL F. BUCHERER

Bucherer's calendar watch features retrograde display of date, second time zone, and power reserve indicator.

The chronograph versions of the Zenith El Primero include a model with retrograde date display.

retrograde displays. Pierre Kunz and, to some extent, Roger Dubuis are offering retrograde displays as unique features, something that is still surprising for most watch buyers, since you rarely set eyes on this complication in everyday life. Patek Philippe enhances its two Grande Complication watches with a retrograde date hand. Vacheron Constantin delivers its Mercator with dials featuring map designs based on the work of cartographer Gerhard Mercator. The watch dials are available showing maps of America, Europe, and Asia. The time is displayed in retrograde style using a golden compass. Anyone looking for a talking piece and who would love to stun those around them, will be well served with a retrograde style display.

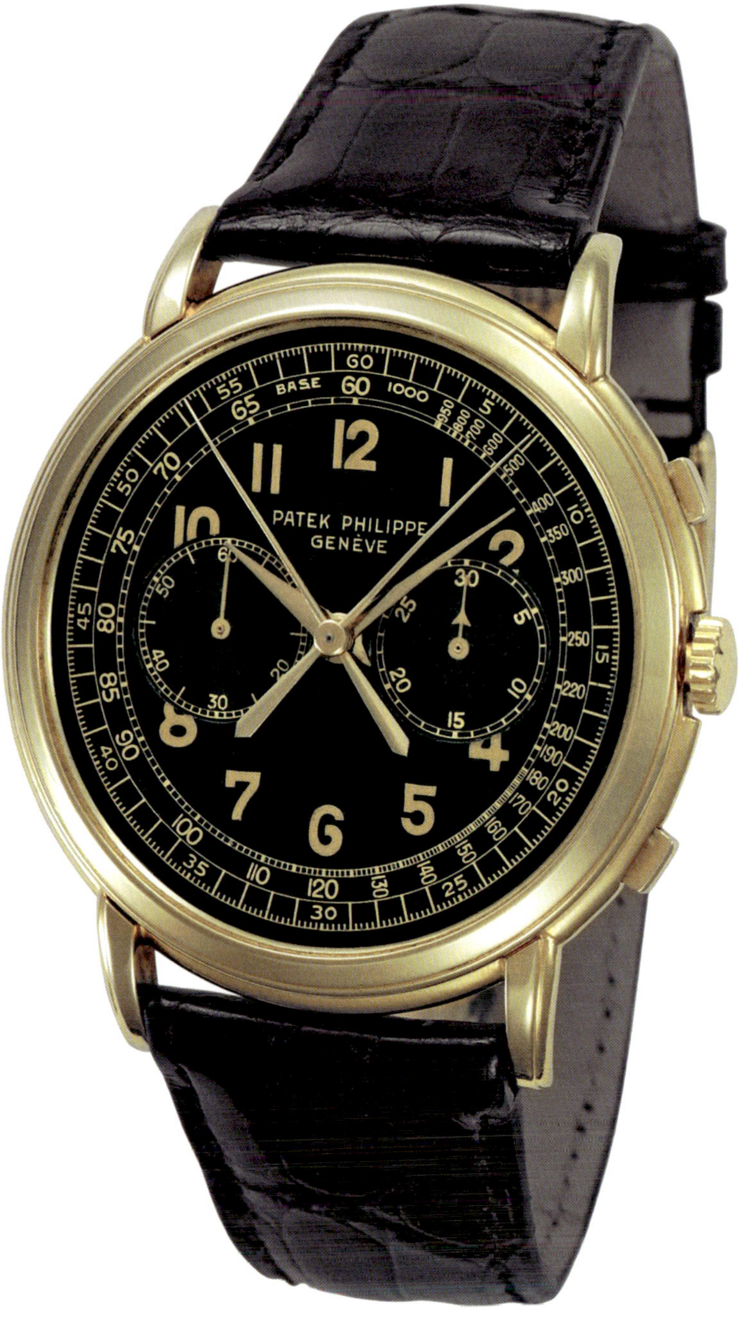
PATEK PHILIPPE
GENÈVE

Vintage Pieces for Your Wrist

Wearing and Collecting Historical Watches

The mechanical watch is an anachronism, although an endearing one. Why not then, logically, refrain from buying a new mechanical watch, and invest your money in a vintage watch? In terms of their functioning capacity, these timepieces have nothing to fear in any comparison with their newer counterparts. Many of the older models are perfect for everyday use, right up to being waterproof; the latter capacity can easily be refurbished, for example, in vintage Rolex Tool Watches.

Many who purchase an old watch don't stop at one. Rather, infected by this new passion, the process begins of hunting at auctions and collectors' fairs for that urgently needed piece for their collection. When choosing a historical or previously owned watch, however, there are some things to consider to avoid unpleasant surprises after the purchase. You should always consider whether this watch fits in with your own collection, and also whether you actually like it. To answer, you always have to ask the first question: What should you actually collect? It would be quite wrong to simply buy watches, according to your taste, that you "only" find pleasing. That may be satisfying for the moment, but in the long run leads to more frustration.

The earlier chapters of this book will also be helpful in terms of historical watches

Pilot's split-seconds chronograph by Patek Philippe Reference 2512 from the year 1950. The third push-piece is operated via the crown. A 30-minute totalizor is at the three.

Very early wristwatch models still invoke their pocket watch origins. A leather strap was simply fitted using the additionally mounted lugs.

and investing in a collection. Complications or brands might be your guidelines. Here, too, further distinctions are possible. For example, regarding chronographs, the collector can limit themselves to manual-w0inding movements, or vice-versa, collect only automatic chronograph calibers. Chronometers are also a popular area for collectors and ensure great brand diversity; as with chronographs, middle-income people can successfully assemble a collection of those pieces. Anyone who wants to collect perpetual calendar watches, however, requires a significantly larger amount of capital, and, in terms of brand, will be dealing with a quite manageable range. Anyone making a particular brand their object of desire, can make careful distinction in their choice. From Rolex, only sports models; from Patek Philippe, only complications: anything is a possibility. There are even those who collect according to the metal used, or their colors: i.e., steel, white gold, platinum, or only yellow or rose gold watches. This collection, built up with your lifeblood and enthusiasm, may yet ultimately prove disappointing in terms of value. While possibly one or another watch may awaken someone's desire, ultimately, no one might want the whole mixed lot.

Current taste often plays a negative role at the time of sale. Today, people are wearing watches of at least 40 mm diameter, so that a more beautiful men's chronograph of just 33 mm is hard to sell. But a 44 mm Zenith Garelli manual-winding chronograph from the 1970s is a sought-after piece at auctions. A mint condition, heavy Glashütte German *Luftwaffe* fighter pilot navigation wristwatch from 1941—assembled by Fischer & Trabrandt in Pforzheim—in the original box, created market interest, with its 55 mm diameter, due to its sheer size. Even someone simply interested in purchasing an attractive watch, who does not belong to the guild of military watch collectors, would be excited about this piece.

Earlier on, watchmakers were ambitious to fit a watch movement into the smallest space possible. As a result, 33-36 mm cases for men's watches were quite common, and even a demonstration of quality. You should also be cautious about any combination of large wristwatches only later fitted with pocket watch movements. No manufacturer will provide service for any such do-it-yourself piece,

Rectangular watches also have a long tradition at IWC. In the 1930s, it was the Caliber 87; the ultimate model was the 1987 Novecento with perpetual calendar.

GRADUÉ POUR 15 PULSATIONS
GRADUÉ POUR 5 RESPIRATIONS
PATEK, PHILIPPE & Co
GENÈVE
REYKJAVICK LONDRES
PARIS GENÈVE
OSLO
RIGA
LE CAIRE
MOSCOU
BAGDAD
KARACHI
DERA
BOMBAY
CALCUTTA
BATAVIA
SHANG-HAI
TOKIO
DARWIN
SYDNEY
WELLINGTON
HONOLULU
ALASKA
MARQUISES
VANCOUVER
DENVER
MEXICO
CHICAGO
NEW YORK
BOGOTA
BUENOS AIRES
RIO DE JANEIRO
ACORES
CANARIES

and would also decline to even just overhaul the movement.

In terms of the brands, it is the classics, such as Audemars Piguet, Breguet, Cartier, Jules Jurgensen, A. Lange & Söhne, Patek Philippe, Rolex, and Vacheron Constantin, which maintain their relative independence from all cycles of value and appeal.

If you customarily also wear your collector's pieces, you have to enjoy wearing a watch on your arm. You can buy at the spur of the moment based on positive vibrations, but then reason has to come in, so as not to be ultimately disappointed. First, you should know the watch's provenance. Where and especially, by whom was it sold? Can you find out who was its first owner? What is the watch's age or what year was it made? Are all parts original? So-called "marriages," which combine a manufacturer's parts, although not as worthless as a simple forgery, do not equal the value of an original collector's piece. The best guarantee of originality is when the appropriate accessories (such as papers, box, and other items; with Rolex, these can include, for example, tissue paper or a small calendar) are available.

If a watch has been given a full overhaul, it is of considerable importance where this

A combination of complications appropriate for world travelers was already available by 1940. This Patek Philippe Reference 1415-1 featuresa world time display as well as a chronograph with tachymeter and pulsometer scale.

repair work was done. For most manufacturers, a major overhaul means restoring the watch to mint condition. It goes without saying that only original spare parts from the appropriate time period should be used. It is helpful for the buyer if the seller can present any documents about repair work.

Among classic watches, one brand is the undisputed leader. Some 50 percent of watches sold at auction worldwide are Patek Philippe pieces. A Reference 2526J HL, a normal three-hand watch with enamel dial, made between 1953-1960, today will achieve a price of some $30,000. In 1978, such a watch could be bought for $900. Relatively new models also increase in value. A Reference 3970E with Caliber CH27-70Q·36, a platinum men's wristwatch with chronograph, 24-hour display, perpetual calendar, leap year display, and moon phase from 1998, cost about $92,500 in 2008. But also important is the condition of the watch and its accessories. In this case, the original mahogany box, the platinum screw-down case back supplied at delivery along with the transparent display back, the setting pin and a Extract from the Patek Archives, all document the watch's authenticity.

Stefan Muser, owner of Auktionshaus Dr. Crott, one of the top addresses for classic pieces, noted about this watch, "... the Ref. 3970 has meanwhile gained the reputation of a true classic, and, in terms of collectors' favor, barely comes behind the famous References 1518, 2499, and 2499/100. Today, only a collector who is as equally well-off as they are patient, would be able

If, as for this Rolex Submariner, the original box plus all the papers and chronometer certificate are available, you can buy with confidence.

to add one of these three References to their collection. So it is not surprising that the Reference 3970 is gaining more and more importance."

You have to be able to invest significantly more for one of the rare chronographs. To obtain a 1953 Reference 1463 18k rose gold chronograph, with a two-piece screw-down case back and a rose gold Patek Philippe clasp, with a Caliber 13''' in a 35 mm diameter case, would cost a quarter million. The Reference 1436 moves in a similar price category; this is a 1939 split-seconds chronograph, with a petite 33 mm diameter. Patek Philippe has been making these special bracelet chronographs since 1923.

Even more expensive is such an extreme rarity as the Reference 1518, of which only two pieces in steel cases are known. In this model, the steel version is much more expensive than the precious metal versions. This chronograph in steel case with perpetual calendar and moon phase, would be, due to its exotic quality, the crowning piece of any Patek Philippe collection. Of course, here we are in the absolute high-end range, something we can often only experience by reading about it. However, even the average

Fanatical collectors will pay top prices for a Patek Philippe Reference 1518 from 1942 with steel case; today, there are still two of these watches. The perpetual calendar with chronograph in gold is significantly less expensive.

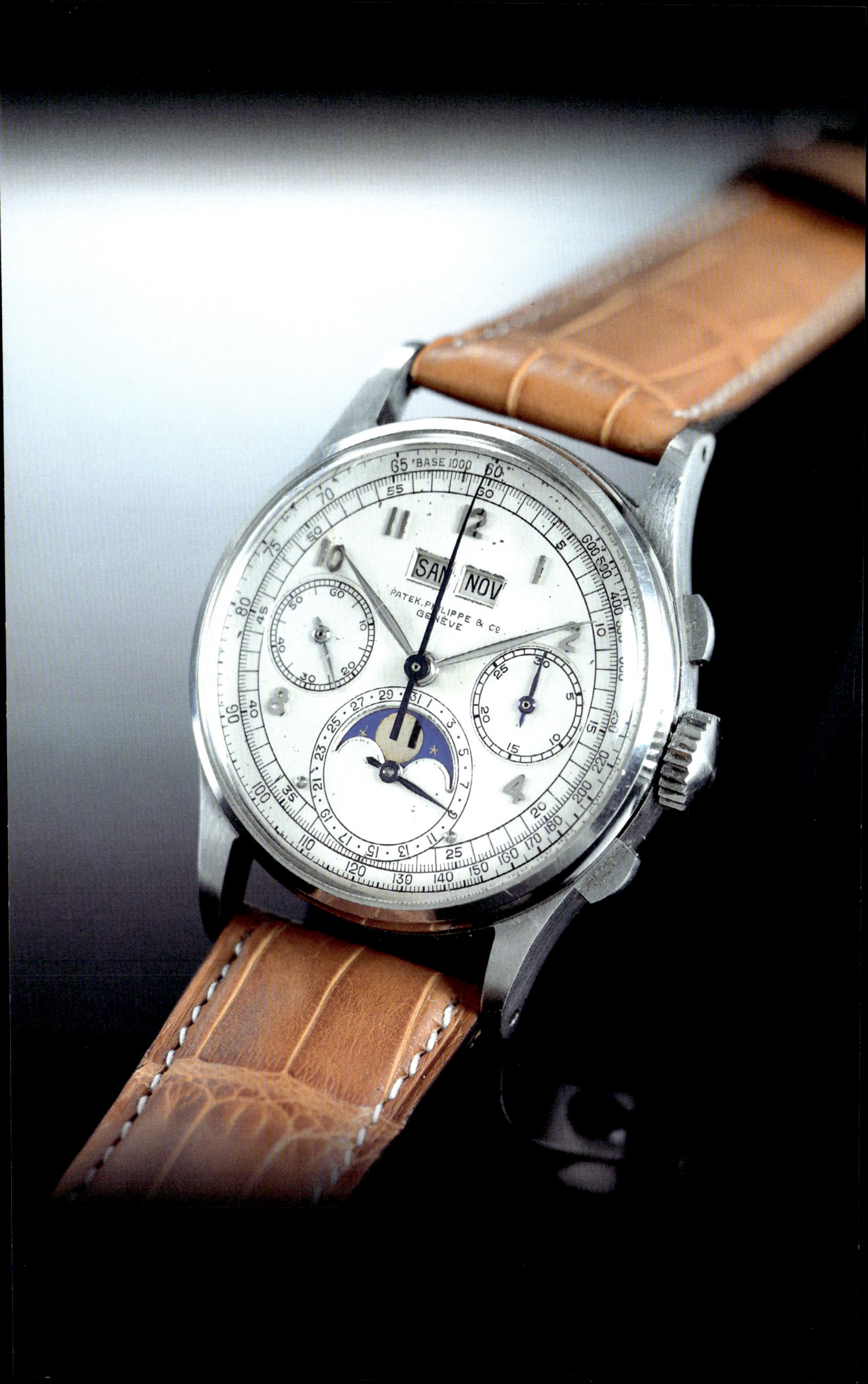
BASE 1000
SAM
NOV
PATEK PHILIPPE & Co
GENÈVE

Audemars Piguet, with the thinnest automatic movement with central rotor. The Caliber 2120 with perpetual calendar module 2800 OP is a feast for the eyes, due to the view allowed by the skeletonized rotor alone.

earner can enjoy indulging themselves in the hobby of watch collecting. Most of the Rolex Tool Watches are still in an affordable range, if the watch does not involve a chronograph with calendar and/or moon phase display. A Submariner or GMT, like an early Milgauss, are watches wonderfully suitable for daily wear and also can be, at times, bought for less than $14,000. An Omega Speedmaster Professional, "the first and only watch worn on the moon" Reference 145.0057, Cal. 1866, with 42 mm diameter, can be had for $4,000 or even less. You can buy the original Moon Watch, without moon phase, with Caliber 321/861 at even lower cost.

Among these interesting watches is the International Watch Co. Schaffhausen's Mark XI. This watch, equipped with the Caliber 89, was frequently used by the *Luftwaffe* and distinguished mainly by its anti-magnetic properties. It was famous for its 648-hour test, which every watch had to pass before delivery. In terms of price, such an IWC is in the range between $3,500-4,000 and can also be purchased in the 1950 model, as a reminder of its birth year. A *Luftwaffe* IWC from 1940 with matt steel case and a diameter of 55 mm is ten times more expensive. Only 1,200 of these watches with Caliber 52 T-19 were delivered to Berlin.

Other interesting military watches include the Junghans *Bundeswehr* chronograph for the *Luftwaffe* with Caliber J 88, or the successor model by Heuer, with its Valjoux 722 which even had a flyback function. Both models are available, depending on their condition, for between $2,000-3,500. The latter, with 44 mm case diameter, exactly suits contemporary taste.

With a little luck, you can also purchase other major brands at reasonable prices. An Audemars Piguet perpetual calendar of Reference 25850OR.002 with Caliber 2120/2802 from a limited series of 50 watches from the year 1998, will set you back only $12,000. This is a real bargain for a rose gold watch with these complications. More expensive, and even only in a steel case, is a model from the oldest watch factory. A Vacheron Constantin chronograph Caliber 434 with a 34-mm case and original folding clasp from 1940, costs significantly more than $40,000. Since a Picard Cadet 1950 chronograph with a Venus 175 Caliber in solid red gold has a similar appearance and costs around $6,700, it can be had for significantly less money, but without less enjoyment. Chronographs from Breitling, Ebel, Breguet, and Heuer, a Cartier Tank—these are all ways to obtain a bargain in the moderate price range up to $4,000. The enjoyment of having a vintage watch can already start at $1,300.

For rectangular watch aficionados, the Jaeger-LeCoultre Reverso, in all its shapes and sizes, offers a more affordable range. An Art Deco with Caliber 823 in good condition with box and papers will cost some $8,000. The Reverso is also available as a tourbillion, but probably more expensive. A rectangular IWC Novecento, which anyway has a perpetual calendar, is also available at these prices.

IWC also makes the least-costly Grande Complication, in gold or platinum case with 42 mm diameter. Reference 3770 has a minute repeater, chronograph, perpetual calendar, and moon phase. At work inside is an extremely complex minute repeater chiming mechanism with all-or-nothing function. This invention signals the time with quiet tones, coming from two gongs that are set off by a slider on the left edge of the case. These are struck by two small precision hammers at every passing hour, quarter-hour, and minute. This chiming mechanism is a technical masterpiece. Because at first the sound could not penetrate the solid precious metal case, the crystal was freely suspended on a metal membrane, amplifying the gong vibrations by the sound transmission pin. Accommodating in terms of expense? Merely some $107,000.

Also in this price category is a simple two-hand watch with an unprepossessing exterior: one of the first Panerai watches manufactured by Rolex. Since this is a presentation-model watch prototype for Reference 3646 with California dial, it commands a price only comprehensible to a fanatical Paneristi. A beautiful Cartier, an

Audemars Piguet Royal Oak with a complication such as chronograph or perpetual calendar—these are watches that provide an outstanding accessory to accompany you through everyday life. The Zenith models of the 1980s and 1990s, such as the El Primero Calibers 400 and 410, especially, are often obtainable at auction at relatively favorable prices. There are a few calendar chronographs with chronometer certificate, mostly associated with the Valjoux 88, which come close to the simple yet high-quality appearance of these models.

It can be stated, ergo: If you are a lover of classic watch designs, you can go least wrong with a vintage watch—assuming it is the right brand.

The new Lange Zeitwerk displays hours and minutes digitally. The precisely jumping time displays are powered by manual winding Caliber L 043.1. which has a power reserve of 36 hours.